The Wireless Tesla

The Wireless Tesla

By Nikola Tesla

Table of Contents

Tesla's Wireless Light

Scientific American — Feb. 2, 1901

Nikola Tesla has given to The New York Sun an authorized statement concerning his new experiments on the production of light without the aid of wires. Mr. Tesla says:

"This light is the result of continuous efforts since my early experimental demonstrations before scientific societies here and abroad. In order to make it suitable for commercial use, I had to overcome great difficulties. One of these was to produce from ordinary currents of supply electrical oscillations of enormous rapidity in a simple and economical manner. This, I am glad to say, I have now accomplished, and the results show that with this new form of light a higher economy is practicable than with the present illuminants. The light offers, besides, many specific advantages, not the least of which is found in its hygienic properties. It is, I believe, the closest approach to daylight which has yet been reached from any artificial source.

"The lamps are glass tubes which may be bent in any ornamental way. I most generally use a rectangular spiral, containing about twenty to twenty-five feet of tubing making some twelve to fourteen convolutions. The total illuminating surface of a lamp is from 300 to 400 square inches. The ends of the spiral tube are covered with a metallic coating, and provided with hooks for hanging the lamp on the terminals of the source of oscillations. The tube contains gases rarefied to a certain degree, determined in the course of long experimentation as being conductive to the best results.

"The process of light production is, according to my views, as follows: The street current is passed through a machine which is an electrical oscillator of peculiar construction and transforms the supply current, be it direct or alternating, into electrical oscillations of very high frequency. These oscillations, coming to the metallically-coated ends of the glass tube, produce in the interior corresponding electrical oscillations, which set the molecules and atoms of the inclosed rarefied gases into violent commotion, causing them to vibrate at enormous rates and emit those radiations which we know as light. The gases are not rendered incandescent in the ordinary sense, for if it were so, they would be hot, like an incandescent filament. As a matter of fact, there is very little heat noticeable, which speaks well for the economy of the light, since all heat would be loss.

"This high economy results chiefly from three causes: First, from the high rate of the electrical oscillations; second, from the fact that the entire light-giving body, being a highly attenuated gas, is exposed and can throw out its radiations unimpeded, and, third, because of the smallness of the particles composing the light-giving body, in consequence of which they can be quickly thrown into a high rate of vibration, so that comparatively little energy is lost in the lower or heat vibrations. An important practical advantage is that the lamps need not be renewed like the ordinary ones, as there is nothing in them to consume. Some of these lamps I have had for years, and they are now in just as good a condition as they ever were. The illuminating power of each of these lamps is, measured by the photometric method, about fifty candle power, but I can make them of any power desired, up to that of several arc lights. It is a remarkable feature of the light that during the day it can scarcely be seen, whereas at night the whole room is brilliantly illuminated. When the eye becomes used to the light of these tubes, an ordinary incandescent lamp or gas burner produces a violent pain in the eye when it is turned on, showing in a striking manner to what a degree these concentrated sources of light which we now use are detrimental to the eye.

"I have found that in almost all its actions the light produces the same effects as-sunlight, and this makes me hopeful that its introduction into dwellings will have the effect of improving, in a measure now impossible to estimate, the hygienic conditions. Since sunlight is a very powerful curative agent, and since this light makes it possible to have sunlight, so to speak, of any desired intensity, day and night in our homes, it stands to reason that the development of germs will be checked and many diseases, as consumption, for instance, successfully combated by continually exposing the patients to the rays of these lamps. I have ascertained unmistakably that the light produces a soothing action on the nerves, which I attribute to the effect which it has upon the retina of the eye. It also improves vision just exactly as the sunlight, and it ozonizes slightly the atmosphere. These effects can be regulated at will. For instance, in hospitals, where such a light is of paramount importance, lamps may be designed which will produce just that quality of ozone which the physician may desire for the purification of the atmosphere, or if necessary, the ozone production can be stopped altogether.

"The lamps are very cheap to manufacture, and by the fact that they need not be exchanged like ordinary lamps or burners they are rendered still less expensive. The chief consideration is, of course, in commercial introduction, the energy consumption. While I am not yet prepared to give exact figures, I can say that, given a certain quantity of electrical energy from the mains, I can produce more light than can be produced by the ordinary methods. In introducing this system of lighting my transformer, or oscillator, will be usually located at some convenient place in the basement, and from there the transformed currents will be led as usual through the building. The lamps can be run with one wire alone, as I have shown in my early demonstrations, and in some cases I can dispense entirely with the wires. I hope that ultimately we shall get to this ideal form of illumination, and that we shall have in our rooms lamps which will be set aglow no matter where they are placed, just as an object is heated by heat rays emanating from a stove. The lamps will then be handled like kerosene lamps, with this difference, however, that the energy will be conveyed through space. The ultimate perfection of apparatus for the production of electrical oscillations will probably bring us to this great realization, and then we shall finally have the light without heat or 'cold' light. I have no difficulty now to illuminate the room with such wireless lamps, but a number of improvements must be made yet before it can be generally introduced."

The Transmission of Electrical Energy Without Wires
Electrical World and Engineer, March 5, 1904

It is impossible to resist your courteous request extended on an occasion of such moment in the life of your journal. Your letter has vivified the memory of our beginning friendship, of the first imperfect attempts and undeserved successes, of kindnesses and misunderstandings. It has brought painfully to my mind the greatness of early expectations, the quick flight of time, and alas! the smallness of realizations. The following lines which, but for your initiative, might not have been given to the world for a long time yet, are an offering in the friendly spirit of old, and my best wishes for your future success accompany them.

Towards the close of 1898 a systematic research, carried on for a number of years with the object of perfecting a method of transmission of electrical energy through the natural medium, led me to recognize three important necessities: First, to develop a transmitter of great power; second, to perfect means for individualizing and isolating the energy transmitted; and, third, to ascertain the laws of propagation of currents through the earth and the atmosphere. Various reasons, not the least of which was the help proffered by my friend Leonard E. Curtis and the Colorado Springs Electric Company, determined me to select for my experimental investigations the large plateau, two thousand meters above sea-level, in the vicinity of that delightful resort, which I reached late in May, 1899. I had not been there but a few days when I congratulated myself on the happy choice and I began the task, for which I had long trained myself, with a grateful sense and full of inspiring hope. The perfect purity of the air, the unequaled beauty of the sky, the imposing sight of a high mountain range, the quiet and restfulness of the place—all around contributed to make the conditions for scientific observations ideal. To this was added the exhilarating influence of a glorious climate and a singular sharpening of the senses. In those regions the organs undergo perceptible physical changes. The eyes assume an extraordinary limpidity, improving vision; the ears dry out and become more susceptible to sound. Objects can be clearly distinguished there at distances such that I prefer to have them told by someone else, and I have heard—this I can venture to vouch for—the claps of thunder seven and eight hundred kilometers away. I might have done better still, had it not been tedious to wait for the sounds to arrive, in definite intervals, as heralded precisely by an electrical indicating apparatus—nearly an hour before.

In the middle of June, while preparations for other work were going on, I arranged one of my receiving transformers with the view of determining in a novel manner, experimentally, the electric potential of the globe and studying its periodic and casual fluctuations. This formed part of a plan carefully mapped out in advance. A highly sensitive, self-restorative device, controlling a recording instrument, was included in the secondary circuit, while the primary was connected to the ground and an elevated terminal of adjustable capacity. The variations of potential gave rise to electric surgings in the primary; these generated secondary currents, which in turn affected the sensitive device and recorder in proportion to their intensity. The earth was found to be, literally, alive with electrical vibrations, and soon I was deeply absorbed in the interesting investigation. No better opportunities for such observations as I intended to make could be found anywhere. Colorado is a country famous for the natural displays of electric force. In that dry and rarefied atmosphere the sun's rays beat the objects with fierce intensity. I raised steam, to a dangerous pressure, in barrels filled with concentrated salt solution, and the tin-foil coatings of some of my elevated terminals shriveled up in the fiery blaze. An experimental high-tension transformer, carelessly exposed to the rays of the setting sun, had most of its insulating compound melted out and was rendered useless. Aided by the dryness and rarefaction of the air, the water evaporates as in a boiler, and static electricity is developed in abundance. Lightning discharges are, accordingly, very frequent and sometimes of inconceivable violence. On one occasion approximately twelve thousand discharges occurred in two hours, and all in a radius of certainly less than fifty kilometers from the laboratory. Many of them resembled gigantic trees of fire with the trunks up or down. I never saw fire balls, but as compensation for my disappointment I succeeded later in determining the mode of their formation and producing them artificially.

In the latter part of the same month I noticed several times that my instruments were affected stronger by discharges taking place at great distances than by those near by. This puzzled me very much. What was the cause? A number of observations proved that it could not be due to the differences in the intensity of the individual discharges, and I readily ascertained that the phenomenon was not the result of a varying relation between the periods of my receiving circuits and those of the terrestrial disturbances. One night, as I was walking home with an assistant, meditating over these experiences, I was suddenly staggered by a thought. Years ago, when I wrote a chapter of my lecture before the Franklin Institute and the National Electric Light Association, it had presented itself to me, but I dismissed it as absurd and impossible. I banished it again. Nevertheless, my instinct was aroused and somehow I felt that I was nearing a great revelation.

It was on the third of July—the date I shall never forget—when I obtained the first decisive experimental evidence of a truth of overwhelming importance for the advancement of humanity. A dense mass of strongly charged clouds gathered in the west and towards the evening a violent storm broke loose which, after spending much of its fury in the mountains, was driven away with great velocity over the plains. Heavy and long persisting arcs formed almost in regular time intervals. My observations were now greatly facilitated and rendered more accurate by the experiences already gained. I was able to handle my instruments quickly and I was prepared. The recording apparatus being properly adjusted, its indications became fainter and fainter with the increasing distance of the storm, until they ceased altogether. I was watching in eager expectation. Surely enough, in a little while the indications again began, grew stronger and stronger and, after passing through a maximum, gradually decreased and ceased once more. Many times, in regularly recurring intervals, the same actions were repeated until the storm which, as evident from simple computations, was moving with nearly constant speed, had retreated to a distance of about three hundred kilometers. Nor did these strange actions stop then, but continued to manifest themselves with undiminished force. Subsequently, similar observations were also made by my assistant, Mr. Fritz Lowenstein, and shortly afterward several admirable opportunities presented themselves which brought out, still more forcibly, and unmistakably, the true nature of the wonderful phenomenon. No doubt, whatever remained: I was observing stationary waves.

As the source of disturbances moved away the receiving circuit came successively upon their nodes and loops. Impossible as it seemed, this planet, despite its vast extent, behaved like a conductor of limited dimensions. The tremendous significance of this fact in the transmission of energy by my system had already become quite clear to me. Not only was it practicable to send telegraphic messages to any distance without wires, as I recognized long ago, but also to impress upon the entire globe the faint modulations of the human voice, far more still, to transmit power, in unlimited amounts, to any terrestrial distance and almost without loss.

With these stupendous possibilities in sight, and the experimental evidence before me that their realization was henceforth merely a question of expert knowledge, patience and skill, I attacked vigorously the development of my magnifying transmitter, now, however, not so much with the original intention of producing one of great power, as with the object of learning how to construct the best one. This is, essentially, a circuit of very high self-induction and small resistance which in its arrangement, mode of excitation and action, may be said to be the diametrical opposite of a transmitting circuit typical of telegraphy by Hertzian or electromagnetic radiations. It is difficult to form an adequate idea of the marvelous power of this unique appliance, by the aid of which the globe will be transformed. The electromagnetic radiations being reduced to an insignificant quantity, and proper conditions of resonance maintained, the circuit acts like an immense pendulum, storing indefinitely the energy of the primary exciting impulses and impressions upon the earth of the primary exciting impulses and impressions upon the earth and its conducting atmosphere uniform harmonic oscillations of intensities which, as actual tests have shown, may be pushed so far as to surpass those attained in the natural displays of static electricity.

Simultaneously with these endeavors, the means of individualization and isolation were gradually improved. Great importance was attached to this, for it was found that simple tuning was not sufficient to meet the vigorous practical requirements. The fundamental idea of employing a number of distinctive elements, co-operatively associated, for the purpose of isolating energy transmitted, I trace directly to my perusal of Spencer's clear and suggestive exposition of the human nerve mechanism. The influence of this principle on the transmission of intelligence, and electrical energy in general, cannot as yet be estimated, for the art is still in the embryonic stage; but many thousands of simultaneous telegraphic and telephonic messages, through one single conducting channel, natural or artificial, and without serious mutual interference, are certainly practicable, while millions are possible. On the other hand, any desired degree of individualization may be secured by the use of a great number of co-operative elements and arbitrary variation of their distinctive features and order of succession. For obvious reasons, the principle will also be valuable in the extension of the distance of transmission.

Progress though of necessity slow was steady and sure, for the objects aimed at were in a direction of my constant study and exercise. It is, therefore, not astonishing that before the end of 1899 I completed the task undertaken and reached the results which I have announced in my article in the Century Magazine of June, 1900, every word of which was carefully weighed.

Much has already been done towards making my system commercially available, in the transmission of energy in small amounts for specific purposes, as well as on an industrial scale. The results attained by me have made my scheme of intelligence transmission, for which the name of "World Telegraphy" has been suggested, easily realizable. It constitutes, I believe, in its principle of operation, means employed and capacities of application, a radical and fruitful departure from what has been done heretofore. I have no doubt that it will prove very efficient in enlightening the masses, particularly in still uncivilized countries and less accessible regions, and that it will add materially to general safety, comfort and convenience, and maintenance of peaceful relations. It involves the employment of a number of plants, all of which are capable of transmitting individualized signals to the uttermost confines of the earth. Each of them will be preferably located near some important center of civilization and the news it receives through any channel will be flashed to all points of the globe. A cheap and simple device, which might be carried in one's pocket, may then be set up somewhere on sea or land, and it will record the world's news or such special messages as may be intended for it. Thus the entire earth will be converted into a huge brain, as it were, capable of response in every one of its parts. Since a single plant of but one hundred horse-power can operate hundreds of millions of instruments, the system will have a virtually infinite working capacity, and it must needs immensely facilitate and cheapen the transmission of intelligence.

The first of these central plants would have been already completed had it not been for unforeseen delays which, fortunately, have nothing to do with its purely technical features. But this loss of time, while vexatious, may, after all, prove to be a blessing in disguise. The best design of which I know has been adopted, and the transmitter will emit a wave complex of total maximum activity of ten million horse-power, one per cent. of which is amply sufficient to "girdle the globe." This enormous rate of energy delivery. approximately twice that of the combined falls of Niagara, is obtainable only by the use of certain artifices, which I shall make known in due course.

For a large part of the work which I have done so far I am indebted to the noble generosity of Mr. J. Pierpont Morgan, which was all the more welcome and stimulating, as it was extended at a time when those, who have since promised most, were the greatest of doubters. I have also to thank my friend, Stanford White, for much unselfish and valuable assistance. This work is now far advanced, and though the results may be tardy, they are sure to come.

Meanwhile, the transmission of energy on an industrial scale is not being neglected. The Canadian Niagara Power Company have offered me a splendid inducement, and next to achieving success for the sake of the art, it will give me the greatest satisfaction to make their concession financially profitable to them. In this first power plant, which I have been designing for a long time, I propose to distribute ten thousand horse-power under a tension of one hundred million volts, which I am now able to produce and handle with safety.

This energy will be collected all over the globe preferably in small amounts, ranging from a fraction of one to a few horse-power. One its chief uses will be the illumination of isolated homes. I takes very little power to light a dwelling with vacuum tubes operated by high-frequency currents and in each instance a terminal a little above the roof will be sufficient. Another valuable application will be the driving of clocks and other such apparatus. These clocks will be exceedingly simple, will require absolutely no attention and will indicate rigorously correct time. The idea of impressing upon the earth American time is fascinating and very likely to become popular. There are innumerable devices of all kinds which are either now employed or can be supplied, and by operating them in this manner I may be able to offer a great convenience to whole world with a plant of no more than ten thousand horse-power. The introduction of this system will give opportunities for invention and manufacture such as have never presented themselves before.

Knowing the far-reaching importance of this first attempt and its effect upon future development, I shall proceed slowly and carefully. Experience has taught me not to assign a term to enterprises the consummation of which is not wholly dependent on my own abilities and exertions. But I am hopeful that these great realizations are not far off, and I know that when this first work is completed they will follow with mathematical certitude.

When the great truth accidentally revealed and experimentally confirmed is fully recognized, that this planet, with all its appalling immensity, is to electric currents virtually no more than a small metal ball and that by this fact many possibilities, each baffling imagination and of incalculable consequence, are rendered absolutely sure of accomplishment; when the first plant is inaugurated and it is shown that a telegraphic message, almost as secret and non-interferable as a thought, can be transmitted to any terrestrial distance, the sound of the human voice, with all its intonations and inflections, faithfully and instantly reproduced at any other point of the globe, the energy of a waterfall made available for supplying light, heat or motive power, anywhere-on sea, or land, or high in the air-humanity will be like an ant heap stirred up with a stick: See the excitement coming!

The Transmission of Electrical Energy Without Wires as a Means for Furthering Peace

Electrical World and Engineer, January 7, 1905, pp. 21-24

Universal Peace, assuming it to be in the fullest sense realizable, might not require eons for its accomplishment, however probable this may appear, judging from the imperceptibly slow growth of all great reformatory ideas of the past. Man, as a mass in movement, is inseparable from sluggishness and persistence in his life manifestations, but it does not follow from this that any passing phase, or any permanent state of his existence, must necessarily be attained through a *stataclitic* process of development.

Our accepted estimates of the duration of natural metamorphoses, or changes in general, have been thrown in doubt of late. The very foundations of science have been shaken. We can no longer believe in the Maxwellian hypothesis of transversal ether-undulations of electrical vibrations, this most important field of human endeavor, particularly in the advancement of philanthropy and peace, was in no small measure retarded by that fascinating illusion, which I since long hoped to dispel. I have noted with satisfaction the first signs of a change of scientific opinion. The brilliant discovery of the exceptionally "radio-active" substances, radium and polonium, by Mrs. Sklodowska Curie, has likewise afforded me much personal gratification, being an eclatant confirmation of my early experimental demonstrations, of electrified radian streams of primary matter or corpuscular emanations (*Electrical Review*, New York, 1896-1897), which were then received with incredulity. They have awakened us from the poetical dream of an intangible conveyor of energy, weightless, structureless ether, to the plain, palpable reality of a ponderous medium of coarse particles, or bodily carriers of force. They have led us to a radically new interpretation of the changes and transformations we perceive. Enlightened by this recognition, we cannot say the sun is hot, the moon is cold, the star is bright, for all these might be purely electrical phenomena. If this be the case, then even our conceptions of time and space may have to be modified.

So, too, as regards the organic world, a similar revolution of thought is distinctly observable. In biological and zoological research the bold ideas of Haensel have found support in recent discoveries. A heretic belief in such possibilities as the artificial production of simple living material aggregates, the spontaneous natural creation of complex organisms and willful sex control, is gaining ground. We still brush it aside, but not with pedantic disdain as before. The fact is - our faith in the orthodox theory of slow evolution is being destroyed!

Thus a state of human life vaguely defined by the term "Universal Peace," while a result of cumulative effort through centuries past, might come into existence quickly, not unlike a crystal suddenly forms in a solution which has been slowly prepared. But just as no effect can precede its cause, so this state can never be brought on by any pact between nations, however solemn. Experience is made before the law is formulated, both are related like cause and effect. So long as we are clearly conscious of the expectation, that peace is to result from such a parliamentary decision, so long have we a conclusive evidence that we are not fit for peace. Only then when we shall feel that such international meetings are mere formal procedures, unnecessary except in so far as they might serve to give definite expression to a common desire, will peace be assured.

To judge from current events we must be, as yet, very distant from that blissful goal. It is true that we are proceeding towards it rapidly. There are abundant signs of this progress everywhere. The race enmities and prejudices are decidedly waning. A recent act of His Excellency, the President of the United States, is significant in this respect. We begin to think cosmically. Our sympathetic feelers reach out into the dim distance. The bacteria of the "Weltschmerz," are upon us. So far, however, universal harmony has been attained only in a single sphere of international relationship. That is the postal service. Its mechanism is working satisfactorily, but - how remote are we still from that scrupulous respect of the sanctity of the mail bag! And how much farther again is the next milestone on the road to peace - an international judicial service equally reliable as the postal!

The coming meeting at the Hague, now indefinitely postponed, can only consider temporary expedients. General disarmament being for the present entirely out of question, a proportionate reduction might be recommended. The safety of any country and of the world's commerce depending not on the absolute, but relative amount of war material, this would be evidently the first reasonable step to take towards universal economy and peace. But it would be a hopeless task to establish an equitable basis of adjustment. Population, naval strength, force of army, commercial importance, water-power, or any other natural resource, actual or prospective, are equally unsatisfactory standards to consider.

In view of this difficulty a measure suggested by Carnegie might be adopted by a few strong countries to scare all the weaker ones into peace. But while for the time being such a course may seem advisable, the beneficial effects of this homeopathic treatment of the martial disease could hardly be lasting. In the first place, a coalition of the leading powers could not fail to create an organized opposition, which might result in a disaster all the greater as it was long deferred. The ultimate falling out of the virtuous, peace-dictating nations, as certain as the law of gravitation, should be all the more reckoned with, as it would be extremely demoralizing. Again, it is by no means sufficient authority.

To conquer by sheer force is becoming harder and harder every day. Defensive is getting continuously the advantage of offensive, as we progress in the satanic science of destruction. The new art of controlling electrically the movements and operations of individualized automata at a distance without wires, will soon enable any country to render its coasts impregnable against all naval attacks. It is to be regretted, in this connection, that my proposal to the United States Navy four years ago, to introduce this invention, did not receive the least encouragement. Also that my offer to Secretary Long to establish telegraphic communication across the Pacific Ocean by my wireless system was thrown in the naval waste basket in Washington, quite *sans facon*. At that time I had already announced in *The Century Magazine* of June, 1900, my successful "girdling" of the globe with electrical impulses (stationary waves), and my "telautomata" had been publicly exhibited. But that was not the fault of the naval officials, for then these inventions of mine were decried as bald, visionary schemes, loudest indeed by those who have since become Croesuses of Promise--in "light" storage batteries, "Ocean" telephony and "transatlantic" wireless telegraphy, yet remained to this day—Sisyphuses of Attainment. Had only a few "telautomatic" torpedoes been constructed and adopted by our navy, the mere moral influence of this would have been powerfully and most beneficially felt in the present Eastern Complication. Not to speak of the advantages which might have been secured through the direct and instantaneous transmission of messages to our distant colonies and scenes of the present barbarous conflicts. Since advancing that principle, I have invented a number of improvements, making it possible to direct such a torpedo, submersible at will, from a distance much greater than the range of the largest gun, with unerring precision, upon the object to be destroyed. What is still more surprising, the operator will not need to see the infernal engine or even know its location, and the enemy will be unable to interfere, in the slightest, with its movements by any electrical means. One of these devil-telautomata will soon be constructed, and I shall bring it to the attention of governments. The development of this art must unavoidably arrest the construction of expensive battleships as well as land fortifications, and revolutionize the means and methods of warfare. The distance at which it can strike, and the destructive power of such a quasi-intelligent machine being for all practical purposes unlimited, the gun, the armor of the battleship and the wall of the fortress, lose their import and significance. One can prophesy with a Daniel's confidence that skilled electricians will settle the battles of the near future. But this is the least. In its effect upon war and peace, electricity offers still much greater and more wonderful possibilities. To stop war by the perfection of engines of destruction alone, might consume centuries and centuries. Other means must be employed to hasten the end. What are these to be? Let us consider.

Fights between individuals, as well as governments and nations, invariably result from misunderstandings in the broadest interpretation of this term. Misunderstandings are always caused by the inability of appreciating one another's point of view. This again is due to the ignorance of those concerned, not so much in their own, as in their mutual fields. The peril of a clash is aggravated by a more or less predominant sense of combativeness, posed by every human being. To resist this inherent fighting tendency the best was is to dispel ignorance of the doings of others by a systematic spread of general knowledge. With this object in view, it is most important to aid exchange of thought and intercourse.

Mutual understanding would be immensely facilitated by the use of one universal tongue. But which shall it be, is the great question. At present it looks as if the English might be adopted as such, though it must be admitted that it is not the most suitable. Each language, of course, excels in some feature. The English lends itself to a terse, forceful expression of facts. The French is precise and finely distinctive. The Italian is probably the most melodious and easiest to learn. The Slavic tongues are very rich in sound but extremely difficult to master. The German is unequaled in the facility it offers for coining and combining words. A practical answer to that momentous question must perforce be found in times to come, for it is manifest that by adopting one common language the onward march of man would be prodigiously quickened. I do not believe that an artificial concoction, like Volapuk, will ever find universal acceptance, however time-saving it might be. That would be contrary to human nature. Languages have grown into our hearts. I rather look to the possibility of a reversion to the old Latin or Greek mother tongues, basing myself in this conclusion on the Spencerian law of rhythm [see Spencer's First Principles]. It seems unfortunate that the English-speaking nations, who are now fittest to rule the world, while endowed with extraordinary energy and practical intelligence, are singularly wanting in linguistic talent.

Next to speech we must consider permanent records of all kinds as a means for disseminating general information, or that knowledge of mutual endeavor which is chiefly conducive to harmony. Here the newspapers play by far the most important part. They are undoubtedly more effective than institutions of learning, libraries, museums and individual correspondence, all combined. The knowledge they convey is, on the whole, superficial and sometimes defective, but it is poured out in a mighty stream that reaches far and wide. Disregarding the force of electrical invention, that of journalism is the greatest in urging peace. Our schools are instrumental, mainly, in the furtherance of special thorough knowledge in our own fields, which is destructive of concordance. A world composed of crass specialists only would be perpetually at war. The diffusion of general knowledge through libraries and similar sources of information is very slow. As to individual correspondence, it is principally useful as an indispensable ingredient of the cement of commercial interest, that most powerful binding material between heterogeneous masses of humanity. It would be hard to overestimate the beneficial influence of the marvelous and precise art of photography, nor can that of other arts or means of recording be ignored. But a simple reflection will show that the peace-making force of all permanent, printed, printed or other records, resides not in themselves. It must be sought elsewhere. This is also true of speech.

Our senses enable us to perceive only a minute portion of the outside world. Our hearing extends to a small distance. Our sight is impeded by intervening bodies and shadows. To know each other we must reach beyond the sphere of our sense perceptions. We must transmit our intelligence, travel, transport the materials and transfer the energies necessary for our existence. Following this thought we now realize, forcibly enough to dispense with argument, that of all other conquests of man, without exception, that which is most desirable, which would be most helpful in the establishment of universal peaceful relations is - the complete *Annihilation of Distance*.

To achieve this wonder, electricity is the one and only means. Inestimable good has already been done by the use of this all powerful agent, the nature of which is still a mystery. Our astonishment at what has been accomplished would be uncontrollable were it not held in check by the expectation of greater miracles to come. That one, the greatest of all, can be viewed in three aspects: *Dissemination of intelligence, transportation, and transmission of power.*

Referring to the first, the present systems of telegraphic and telephonic communication are very limited in scope. The conducting channels are costly and of small working capacity. There is serious inductive disturbance, and storms render the service unsafe which, moreover, is too expensive: A vast improvement will be effected by placing the wires underground and insulating them artificially, by refrigeration. Their working capacity also could by indefinitely augmented by resorting to the new principle of "individualization," which I have more recently announced, permitting the simultaneous transmission of thousands of telegraphic and telephonic messages, without interference, over a single wire. The public would be already profiting from these great advances were it not for the stolid indifference of the leading companies engaged in the transmission of intelligence. But new concerns are springing into existence and the near future will witness a great transformation along these two lines of invention. The submarine cables are subject to still greater limitations. Some obstacles to rapid signaling, through them, seem insuperable. The attempts to overcome these have been numerous, but so far all have proved futile. The celebrated mathematician, O. Heaviside, and several able electricians following in his footsteps, have fallen into the singular error that rapid telegraphy and even telephony through ocean cables would be made practicable by the use of induction coils. Inductances might be to some extent helpful on comparatively short lines with thick paper insulation; on long lines insulated with rubber or gutta-percha they would be positively detrimental. Improvements will, undoubtedly, be made, but great electrostatic capacity and unavoidable loss of energy in the insulation and surrounding conductors will always restrict the usefulness of the transmission through artificial conductors is necessarily confined to a small number of stations.

It is therefore evident that the abolishment of all these drawbacks by the conveyance of signals or messages without wires, as I have undertaken in my "world" telegraphy and telephony, will be of the greatest moment in the furtherance of peace. The unifying influence of this advance will be felt all the more, as it will not only completely annihilate distance, but also make it possible to operate from a singe "world" telegraphy plant, an unlimited number of receiving stations distributed all over the globe, and with equal facility, irrespective of location. Within a few years a simple and inexpensive device, readily carried about, will enable one to receive on land or sea the principal news, to hear a speech, a lecture, a song or play of a musical instrument, conveyed from any other region of the globe. The invention will also meet the crying need for cheap transmission to great distances, more especially over the oceans. The small working capacity of the cables and the excessive cost of messages are now fatal impediments in the dissemination of intelligence which can only be removed by transmission without wires.

The deficiencies of Hertzian telegraphy have created in the public mind the impression that exclusive or private messages without the use of artificial channels are impracticable. As a matter of fact, nothing could be more erroneous. Ever since its first appearance in 1891, I have denied the commercial possibilities of the system of signaling by Hertzian or electromagnetic waves, and my forecasts have been fully confirmed. It lends itself little to tuning, still less to the higher artifices of "individualization," and transmission to considerable distances is wholly out of the question. Portentous claims for this method of communication were made three years ago, but they have been unable to stand the hard, cruel test of time. Moreover, I have recently learned through the leading British electrical journal (*Electrician*, London, February 27, 1903), that some experimenters have abandoned all their own and have been "converted" to my methods and appliances, without my approval and officiation. I was both astonished and pained - astonished at the nonchalance and lack of appreciation of these men, pained at the inability exhibited in the construction and use of my apparatus. My high hopes raised by that excellent journal, however, are still to be realized, for I have ascertained that His Majesty the King of England, His excellency the President of the United States, and other persons of exalted positions have, after all, not conferred upon me the imperishable honor of graciously condescending to the use for my coils, transformers and high-potential methods of transmission, but have exchanged their August greetings through the medium of a cable in the old-fashioned way. What has been actually achieved by Hertzian telegraphy can only be conjectured.

Quite different conditions exist in my system in which the electromagnetic waves or radiations are designedly minimized, the connection of one of the terminals of the transmitting circuit to the ground having, itself, the effect of reducing the energy of these radiations to about one-half. Under observance of proper rules and artifices the distance is of little or no consequence, and by skillful application of the principle of "individualization," repeatedly referred to the messages may be rendered both non-interfering and non-interferable. This invention, which I have described in technical publications, attempts to imitate, in a very crude way, the nervous system in the human body. It was the outcome of long-continued tests demonstrating the impossibility of satisfying rigorous commercial requirements by my earlier system, based on simple tuning, in which the selective quality is dependent on a single characteristic feature. In this later improvement the exclusiveness and non-interferability of impulses transmitted through a common channel result from cooperative association for a number of distinctive elements, and can be pushed as far as desired. In actual practice it is found that by combining only two vibrations or tones, a degree of privacy sufficient for most purposes is attained. When three vibrations are combined, it is extremely difficult, even for a skilled expert, to read or to disturb signals not intended for him, with four it is a vain undertaking. The probability of his getting the secret combinations at the right moments and in proper order, is much smaller than that of drawing an ambo, terno or quaterno, respectively, in a lottery. From experimental facts, I conclude that the invention will permit the simultaneous transmission of several millions of separately distinguishable messages through the earth, which, strangely enough, is in this respect much superior to an artificial conductor. This number ought to be sufficient to meet all the pressing necessities of intelligence transmission for at least one century to come. It is important to observe that but on "world" telegraphy plant, such as I am now completing, will have a greater working capacity than all the ocean cables combined. Once these facts are recognized this new art, which I am inaugurating, will sweep the world with the force of a uragan.

In transportation a great change is now going on. The trolley lines are being extended, the steam locomotive is making place for the electric motor. The ocean liners are adopting the turbine. Land travel is being improved by the automobile. The waterfalls are being harnessed and the energy used in the propulsion of cars. The advantages of first generating electricity by a prime mover, are being more and more appreciated. To the majority, this may appear a roundabout way of doing, but in reality it is as direct as the driving of a pulley from another by a belt. The idea is already being applied to railroads, and automobiles of this new type are making their appearance. The ocean vessels are bound to follow. An immense and virgin field will be thus opened up to the manufacturers of electric machinery. Effort towards saving time and money is characteristic of all modern methods of transportation. In many of these new developments, the artificial insulation of the high-tension mains by refrigeration will be very useful. However paradoxical, it is true, that by the use of this invention, power for all industrial purposes can be transmitted to distances of many hundreds of miles, not only without any loss, but with appreciable gain of energy. This is due to the fact that the conductor is much colder than the surrounding medium. The operativeness of this method is restricted to the use of a gaseous refrigerant, no known liquid permitting the attainment of a sufficiently low temperature of the transmission line. Hydrogen is by far the best cooling agent to employ. By its use electric railways can be extended to any desired distance. Owing to the smallness of ohmic loss, the objections to the multiphase system disappear and induction motors with closed coil armatures can be adopted. I find that even transmission through a submarine cable, as from Sweden to England, of great amounts of power is perfectly practicable. But the ideal solution of the problem of transportation will be arrived at only when the complete annihilation of distance in the transmission of power in large amounts shall have become a commercial reality. That day we shall invade the domain of the bird. When the vexing problem of aerial navigation, which has defied his attempts for ages, is solved, man will advance with giant strides.

That electrical energy can be economically transmitted without wires to any terrestrial distance, I have unmistakably established in numerous observations, experiments and measurements, qualitative and quantitative. These have demonstrated that is practicable to distribute power from a central plant in unlimited amounts, with a loss not exceeding a small fraction of one per cent, in the transmission, even to the greatest distance, twelve thousand miles - to the opposite end of the globe. This seemingly impossible feat can now be readily performed by any electrician familiar with the design and construction of my "high-potential magnifying transmitter," the most marvelous electrical apparatus of which I have knowledge, enabling the production of effects of unlimited intensities in the earth and its ambient atmosphere. It is, essentially, a freely vibrating secondary circuit of definite length, very high self-induction and small resistance, which has one of its terminals in intimate direct or inductive connection with the ground and the other with an elevated conductor, and upon which the electrical oscillations of a primary or exciting circuit are impressed under conditions of resonance. To give an idea of the capabilities of this wonderful appliance, I may state that I have obtained, by its means, spark discharges extending through more than one hundred feet and carrying currents of one thousand amperes, electromotive forces approximating twenty million volts, chemically active streamers covering areas of several thousand square feet, and electrical disturbances in the natural media surpassing those caused by lightning, in intensity.

Whatever the future may bring, the universal application of these great principles is fully assured, though it may be long in coming. With the opening of the first power plant, incredulity will give way to wonderment, and this to ingratitude, as ever before. The time is not distant when the energy of falling water will be man's life energy. So far only about million horse-power have been harnessed by my system of alternating-current transmission. This is little, but corresponds, nevertheless to the adding of sixty million indefatigable laborers, working virtually without food and pay, to the world's population. The projects which have come to my own attention, however, contemplate the exploitation of water powers aggregating something like one hundred and fifty million horse-power. Should they be carried out in a quarter of a century, as seems probable from present indications, there will be, on the average two such untiring laborers for every individual. Long before this consummation, coal and oil must cease to be important factors in the sustenance of human life on this planet. It should be borne in mind that electrical energy obtained by harnessing a waterfall is probably fifty times more effective than fuel energy. Since this is the most perfect way of rendering the sun's energy available, the direction of the future material development of man is clearly indicated. He will live on "white coal." Like a babe to the mothers's breast will he cling to his waterfall. "Give us our daily waterfall," will be the prayer of the coming generations. *Deus futurus est deus aquae deiectus!*

But the fact that stationary waves are producible in the earth is of special and, in many ways, still greater significance in the intellectual development of humanity. Popularly explained, such a wave is a phenomenon generically akin to an echo - a result of reflection. It affords a positive and incontrovertible experimental evidence that the electric current, after passing into the earth travels to the diametrically opposite region of the same and rebounding from there, returns to its point of departure with virtually undiminished force. The outgoing and returning currents clash and form nodes and loops similar to those observable on a vibrating cord. To traverse the entire distance of about twenty-five thousand miles, equal to the circumference of the globe, the current requires a certain time interval, which I have approximately ascertained. In yielding this knowledge, nature has revealed one of its most precious secrets, of inestimable consequence to man. So astounding are the facts in this connection, that it would seem as though the Creator, himself, had electrically designed this planet just for the purpose of enabling us to achieve wonders which, before my discovery, could not have been conceived by the wildest imagination. A full account of my discoveries and improvements will be given to the world in a special work which I am preparing. In so far, however, as they relate to industrial and commercial uses, they will be disclosed in patent specifications most carefully drawn.

As stated in a recent article (*Electrical World and Engineer*, March 5, 1904), I have been since some time at work on designs of a power plant which is to transmit ten thousand horse-power without wires. The energy is to be collected al over the earth at many places and in varying amounts.

It should not be understood that the practical realization of this undertaking is necessarily far off. The plans could be easily finished this winter, and if some preliminary work on the foundations could be done in the meantime the plant might be ready for operation before the close of next fall. We would then have at our disposal a unique and invaluable machine. Just this one oscillator would advance the world a century. Its civilizing influence would be felt even by the humblest dweller in the wilderness. Millions of instruments of all kinds, for all imaginable purposes, could be operated from that one machine. Universal time could be distributed by simple inexpensive clocks requiring no attention and running with nearly mathematical precision. Stock-tickers, synchronous movements and innumerable devices of this character could be worked in unison all over the earth. Instruments might be provided for indicating the course of a vessel at sea, the distance traversed, the speed, the hour at any particular place, the latitude and longitude. Incalculable commercial advantages could be thus secured and countless accidents and disasters avoided. Here and there a house might be lighted or some other work requiring a few horse-power performed. What is far more important than this, flying machines might be driven in any part of the world. They could be made to travel swiftly because of their small weight and great motive power. My intention would be to utilize this first plant rather as means of enlightenment, to collect its power in very small amounts, and at as many places as possible. The knowledge that there is throbbing through the earth energy readily available everywhere, would exert a strong stimulus on students, mechanics and inventors of all countries. This would be productive of infinite good. Manufacture would receive a fresh and powerful incentive. Conditions, such as never existed before in commerce, would be brought about. Supply would be ever inadequate to demand. The industries of iron, copper, aluminum, insulated wire and many others, could not fail to derive great and lasting benefits from this development.

The economic transmission of power without wires is of all-surpassing importance to man. By its means he will gain complete mastery of the air, the sea and the desert. It will enable him to dispense with the necessity of mining, pumping, transporting and burning fuel, and so do away with innumerable causes of sinful waste. By its means, he will obtain at any place and in any desired amount, the energy of remote waterfalls - to drive his machinery, to construct his canals, tunnels and highways, to manufacture the materials of his want, his clothing and food, to heat and light his home - year in, year out, ever and ever, by day and by night. It will make the living glorious sun his obedient, toiling slave. It will bring peace and harmony on earth.

Over five years have elapsed since that providential lightning storm on the 3d of July, 1899, of which I told in the article before mentioned, and through which I discovered the terrestrial stationary waves; nearly five years since I performed the great experiment which on that unforgettable day, the dark God of Thunder mercifully showed me in his vast, awe-sounding laboratory. I thought then that it would take a year to establish commercially my wireless girdle around the world. Alas! my first "world telegraphy" plant is not yet completed, its construction has progressed but slowly during the past two years. And this machine I am building is but a plaything, an oscillator of a maximum activity of only ten million horse-power, just enough to throw this planet into feeble tremors, by sign and word - to telegraph and to telephone. When shall I see completed that first power plant, that big oscillator which I am designing! From which a current stronger than that of a welding machine, under a tension of one hundred million volts, is to rush through the earth! Which will deliver energy at the rate of one thousand million horse-power - one hundred Falls of Niagara combined in one, striking the universe with blows - blows that will wake from their slumber the sleepiest electricians, if there be any, on Venus or Mars! . . . It is not a dream, *it is a simple feat of scientific electrical engineering*, only expensive - blind, faint-hearted, doubting world! . . . Humanity is not yet sufficiently advanced to be willingly led by the discover's keen searching sense. But who knows? Perhaps it is better in this present world of ours that a revolutionary idea or invention instead of being helped and patted, be hampered and ill-treated in its adolescence - by want of means, by selfish interest, pedantry, stupidity and ignorance; that it be attacked and stifled; that it pass through bitter trials and tribulations, through the heartless strife of commercial existence. So do we get our light. So all that was great in the past was ridiculed, condemned, combated, suppressed — only to emerge all the more powerfully, all the more triumphantly from the struggle.

Tesla's Wireless Torpedo
New York Times — March 20, 1907

To the Editor of The *New York Times*:

A report in the *Times* of this Morning says that I have attained no practical results with my dirigible wireless torpedo. This statement should be qualified. I have constructed such machines, and shown them in operation on frequent occasions. They have worked perfectly, and everybody who saw them was amazed at their performance.

It is true that my efforts to have this novel means for attack and defense adopted by our Government have been unsuccessful, but this is no discredit to my invention. I have spent years in fruitless endeavor before the world recognized the value of my rotating field discoveries which are now universally applied. The time is not yet ripe for the telautomatic art. If its possibilities were appreciated the nations would not be building large battleships. Such a floating fortress may be safe against an ordinary torpedo, but would be helpless in a battle with a machine which carries twenty tons of explosive, moves swiftly under water, and is controlled with precision by an operator beyond the range of the largest gun.

As to projecting wave-energy to any particular region of the globe, I have given a clear description of the means in technical publications. Not only can this be done by the use of my devices, but the spot at which the desired effect is to be produced can be calculated very closely, assuming the accepted terrestrial measurements to be correct. This, of course, is not the case. Up to this day we do not know a diameter of the globe within one thousand feet. My wireless plant will enable me to determine it within fifty feet or less, when it will be possible to rectify many geodetical data and make such calculations as those referred to with greater accuracy.

Nikola Tesla

New York, March 19, 1907

Wireless on Railroads

New York Times — March 26, 1907

To the Editor of The New York Times:

No argument is needed to show that the railroads offer opportunities for advantageous use of a practical wireless system. Without question, its widest field of application is the conveyance to the trains of such general information as is indispensable for keeping the traveler in touch with the world. In the near future a telegraphic printer of news, a stock ticker, a telephone, and other kindred appliances will form parts of the regular wireless equipment of a railroad train. Success in this sphere is all the more certain, as the new is not antagonistic, but, on the contrary, very helpful to the old. The technical difficulties are minimized by the employment of a transmitter the effectiveness of which is unimpaired by distance.

In view of the great losses of life and property, improved safety devices on the cars are urgently needed. But upon careful investigation it will be found that the outlook in this direction is not very promising for the wireless art. In the first place the railroads are rapidly changing to electric motive power, and in all such cases the lines become available for the operation of all sorts of signaling apparatus, of which she telephone is by far the most important. This valuable improvement is due to Prof. J. Paley, who introduced it in Germany eight years ago. By enabling the engineer or conductor of any train to call up any other train or station along the track and obtain full and unmistakable information, the liability of collisions and other accidents will be greatly reduced. Public opinion should compel the immediate adoption of this invention.

Those roads which do not contemplate this transformation might avail themselves of wireless transmission for similar purposes, but inasmuch as every train will require in addition to a complete outfit an expert operator, many roads may prefer to use a wire, unless a wireless telephone can be offered to them.

Nikola Tesla

New York, March 25, 1907

Mr. Tesla on the Wireless Transmission of Power
N. Y. World May 19, 1907

To the Editor of The World:

I have enjoyed very much the odd prediction of Sir Hugh Bell, President of the Iron and Steel Institute, with reference to the wireless transmission of power, reported in The World of the 10th inst.

With all the respect due to that great institution I would take the liberty to remark that if its President is a genuine prophet he must have overslept himself a trifle. Sir Hugh would honor me if he would carefully peruse my British patent No. 8,200, in which I have recorded some of my discoveries and experiments, and which may influence him to considerably reduce his conservative estimate of one hundred years for the fulfillment of his prophecy.

Personally, basing myself on the knowledge of this art to which I have devoted my best energies, I do not hesitate to state here for future reference and as a test of accuracy of my scientific forecast that flying machines and ships propelled by electricity transmitted without wire will have ceased to be a wonder in ten years from now. I would say five were it not that there is such a thing as "inertia of human opinion" resisting revolutionary ideas.

It is idle to believe that because man is endowed with higher attributes his material evolution is governed by other than general physical laws. If the genius of invention were to reveal to-morrow the secret of immortality, of eternal beauty and youth, for which all humanity is aching, the same inexorable agents which prevent a mass from changing suddenly its velocity would likewise resist the force of the new knowledge until time gradually modifies human thought.

What has amused me still more, however, is the curious interview with Lewis Nixon, the naval contractor, printed in the World of the 11th inst. Is it possible that the famous designer of the Oregon is not better versed in editorial matters than some of my farming neighbors of Shoreham? One cannot escape that conviction.

We are not in the dark as regards the electrical energy contained in the earth. It is altogether too insignificant for any industrial use. The current circulating through the globe is of enormous volume but of small tension, and could perform but little work. Beside, how does Nixon propose to coax the current from the natural path of low resistance into an artificial channel of high resistance? Surely he knows that water does not flow up hill. It is absurd of him to compare the inexhaustible dynamic energy of wind with the magnetic energy of the earth, which is minute in amount and in a static condition.

The torpedo he proposes to build is not novel. The principle is old. I could refer him to some of my own suggestions of nine years ago. There are many practical difficulties in the carrying out of the idea, and as much better means for destroying a submarine are available it is doubtful that such a torpedo will ever be constructed.

Nixon has failed to grasp that in my wireless system the effect does not diminish with distance. The Hertz waves have nothing to do with it except that some of my apparatus may be used in their production. So too a Kohinoor might be employed to cut window-glass. And yet, the seeming paradox can be easily understood by any man of ordinary intelligence.

Imagine only that the earth were a hollow shell or reservoir in which the transmitter would compress some fluid, as air, for operating machinery in various localities. What difference would it make when this reservoir is tapped to supply the compressed fluid to the motor? None whatever, for the pressure is the same everywhere. This is also true of my electrical system, with all considerations in its favor. In such a mechanical system of power distribution great losses are unavoidable and definite limits in the quality of the energy transmitted exist. Not so in the electrical wireless supply. It would not be difficult to convey to one of our liners, say, 50,000 horsepower from a plant located at Niagara, Victoria or other waterfall, absolutely irrespective of location. In fact, there would not be a difference of more than a small fraction of one per cent, whether the source of energy be in the vicinity of the vessel or 12,000 miles away, at the antipodes.

Nikola Tesla
New York, May 16, 1907.

Possibilities of "Wireless"
New York Times — Oct. 22, 1907

To the Editor of The New York Times:
In your issue of the 19th inst. Edison makes statements which cannot fail to create erroneous impressions.

There is a vast difference between primitive Hertzwave signaling, practicable to but a few miles, and the great art of wireless transmission of energy, which enables an expert to transmit, to any distance, not only signals, but power in unlimited amounts, and of which the experiments across the Atlantic are a crude application. The plants are quite inefficient, unsuitable for finer work, and totally doomed to an effect less than one percent of that I attained in my test in 1899.

Edison thinks that Sir Hiram Maxim is blowing hot air. The fact is my Long Island plant will transmit almost its entire energy to the antipodes, if desired. As to Martin's communication I can only say, that I shall be able to attain a wave activity of 800,000,000 horse power and a simple calculation will show, that the inhabitants of that planet, if there be any, need not have a Lord Raleigh to detect the disturbance.

Referring to your editorial comment of even date, the question of wireless interference is puzzling only because of its novelty. The underlying principle is old, and it has presented itself for consideration in numerous forms. Just now it appears in the novel aspects of aerial navigation and wireless transmission. Every human effort must of necessity create a disturbance. What difference is there in essence, between the commotion produced by any revolutionary idea or improvement and that of a wireless transmitter? The specter of interference has been conjured by Hertzwave or radio telegraphy in which attunement is absolutely impossible, simply because the effect diminishes rapidly with distance. But to my system of energy transmission, based on the use of impulses not sensibly diminishing with distance, perfect attunement and the higher artifice of individualization are practicable. As ever, the ghost will vanish with the wireless dawn.

Nikola Tesla
New York, Oct. 21, 1907.

Tesla on Wireless

New York Daily Tribune — Oct. 25, 1907

To the Editor of The Tribune:

Sir: In so far as wireless art is concerned there is a vast difference between the great inventor Thomas A. Edison and myself, integrally in my favor. Mr. Edison knows little of the theory and practice of electrical vibrations; I have, in this special field, probably more experience than any of my contemporaries. That you are not as yet able to impart your wisdom by wireless telephone to some subscriber in any other part of the world, however remote, and that the presses of your valuable paper are not operated by wireless power is largely due to your own effort and those of some of your distinguished confreres of this city, and to the efficient assistance you have received from my celebrated colleagues, Thomas A. Edison and Michael Pupin, assistant consulting wireless engineers. But it was all welcome to me. Difficulty develops resource.

The transmission across the Atlantic was not made by any device of Mr. Marconi's, but by my system of wireless transmission of energy, and I have already given notice by cable to my friend Sir James DeWar and the Royal Institution of this fact. I shall also request some eminent man of science to take careful note of the whole apparatus, its mode of operation, dimensions, linear and electrical, all constants and qualitative performance, so as to make possible its exact reproduction and repetition of the experiments. This request is entirely impersonal. I am a citizen of the United States, and I know that the time will come when my busy fellow citizens, too absorbed in commercial pursuits to think of posterity, will honor my memory. A measurement of the time interval taken in the passage of the signal necessary to the full and positive demonstration will show that the current crosses the ocean with a mean speed of 625,000 miles a second.

The Marconi plants are inefficient, and do not lend themselves to the practice of two discoveries of mine, the "art of individualization," that makes the message non-interfering and non-interferable, and the "stationary waves," which annihilate distance absolutely and make the whole earth equivalent to a conductor devoid of resistance. Were it not for this deficiency, the number of words per minute could be increased at will by "individualizing."

You have already commented upon this advance in terms which have caused me no small astonishment, in view of your normal attitude. The underlying principle is to combine a number of vibrations, preferably slightly displaced, to reduce further the danger of interference, active and passive, and to make the operation of the receiver dependent on the co-operative effect of a number of attuned elements. Just to illustrate what can be done, suppose that only four vibrations were isolated on each transmitter. let those on one side be respectively a, b, c, and d. Then the following individualized lines would be ab, ac, ad, bc, bd, cd, abc, abd, acd, bed and abed. The same article on the other side will give similar combinations, and both together twenty-two lines, which can be simultaneously operated. To transmit one thousand words a minute, only forty-six words on each combination are necessary. If the plants were suitable, not ten years, as Edison thinks, but ten hours would be necessary to put this improvement into practice. To do this Marconi would have to construct the plants, and it will then be observed that the indefatigable Italian has departed from universal engineering customs for the fourth time.

Nikola Tesla

New York, Oct. 24, 1907

My Apparatus, Says Tesla
New York Times — Dec. 20, 1907

To the Editor of the New York Times:

I have read with great interest the report in your issue of to-day that the Danish engineer, Waldemar Poulson, the inventor of the interesting device known as the "telegraphone," has succeeded in transmitting accurately wireless telephonic messages over a distance of 240 miles.

I have looked up the description of the apparatus he has employed in the experiment and find that it comprises:

(1) My grounded resonant transmitting circuit;

(2) my inductive exciter;

(3) the so-called "Tesla transformer";

(4) my inductive coils for raising the tension on the condenser;

(5) my entire apparatus for producing undamped or continuous oscillations;

(6) my concatenated tuned transforming circuits;

(7) my grounded resonant receiving transformer;

(8) my secondary receiving transformer.

I note other improvements of mine, but those mentioned will be sufficient to show that Denmark is a land of easy invention.

The claim that transatlantic wireless telephone service will soon be established by these means is a modest one. To my system distance has absolutely no significance. My own wireless plant will transmit speech across the Pacific with the same precision and accuracy as across the table.

Nikola Tesla

New York, Dec. 19, 1907

The Future of the Wireless Art

Wireless Telegraphy & Telephony — by Walter W. Massie & Charles R. Underhill, 1908

Mr. Nikola Tesla, in a recent interview by the authors, as to the future of the Wireless Art, volunteered the following statement which is herewith produced in his own words.

"A mass in movement resists change of direction. So does the world oppose a new idea. It takes time to make up the minds to its value and importance. Ignorance, prejudice and inertia of the old retard its early progress. It is discredited by insincere exponents and selfish exploiters. It is attacked and condemned by its enemies. Eventually, though, all barriers are thrown down, and it spreads like fire. This will also prove true of the wireless art.

"The practical applications of this revolutionary principle have only begun. So far they have been confined to the use of oscillations which are quickly damped out in their passage through the medium. Still, even this has commanded universal attention. What will be achieved by waves which do not diminish with distance, baffles comprehension.

"It is difficult for a layman to grasp how an electric current can be propagated to distances of thousands of miles without diminution of intention. But it is simple after all. Distance is only a relative conception, a reflection in the mind of physical limitation. A view of electrical phenomena must be free of this delusive impression. However surprising, it is a fact that a sphere of the size of a little marble offers a greater impediment to the passage of a current than the whole earth. Every experiment, then, which can be performed with such a small sphere can likewise be carried out, and much more perfectly, with the immense globe on which we live. This is not merely a theory, but a truth established in numerous and carefully conducted experiments. When the earth is struck mechanically, as is the case in some powerful terrestrial upheaval, it vibrates like a bell, its period being measured in hours. When it is struck electrically, the charge oscillates, approximately, twelve times a second. By impressing upon it current waves of certain lengths, definitely related to its diameter, the globe is thrown into resonant vibration like a wire, stationary waves forming, the nodal and ventral regions of which can be located with mathematical precision. Owing to this fact and the spheroidal shape of the earth, numerous geodetical and other data, very accurate and of the greatest scientific and practical value, can be readily secured. Through the observation of these astonishing phenomena we shall soon be able to determine the exact diameter of the planet, its configuration and volume, the extent of its elevations and depressions, and to measure, with great precision and with nothing more than an electrical device, all terrestrial distances. In the densest fog or darkness of night, without a compass or other instruments of orientation, or a timepiece, it will be possible to guide a vessel along the shortest or orthodromic path, to instantly read the latitude and longitude, the hour, the distance from any point, and the true speed and direction of movement. By proper use of such disturbances a wave may be made to travel over the earth's surface with any velocity desired, and an electrical effect produced at any spot which can be selected at will and the geographical position of which can be closely ascertained from simple rules of trigonometry.

"This mode of conveying electrical energy to a distance is not 'wireless' in the popular sense, but a transmission through a conductor, and one which is incomparably more perfect than any artificial one. All impediments of conduction arise from confinement of the electric and magnetic fluxes to narrow channels. The globe is free of such cramping and hinderment. It is an ideal conductor because of its immensity, isolation in space, and geometrical form. Its singleness is only an apparent limitation, for by impressing upon it numerous non-interfering vibrations, the flow of energy may be directed through any number of paths which, though bodily connected, are yet perfectly distinct and separate like ever so many cables. Any apparatus, then, which can be operated through one or more wires, at distances obviously limited, can likewise be worked without artificial conductors, and with the same facility and precision, at distances without limit other than that imposed by the physical dimensions of the globe.

"It is intended to give practical demonstrations of these principles with the plant illustrated. As soon as completed, it will be possible for a business man in New York to dictate instructions, and have them instantly appear in type at his office in London or elsewhere. He will be able to call up, from his desk, and talk to any telephone subscriber on the globe, without any change whatever in the existing equipment. An inexpensive instrument, not bigger than a watch, will enable its bearer to hear anywhere, on sea or land, music or song, the speech of a political leader, the address of an eminent man of science, or the sermon of an eloquent clergyman, delivered in some other place, however distant. In the same manner any picture, character, drawing, or print can be transferred from one to another place. Millions of such instruments can be operated from but one plant of this kind. More important than all of this, however, will be the transmission of power, without wires, which will be shown on a scale large enough to carry conviction. These few indications will be sufficient to show that the wireless art offers greater. possibilities than any invention or discovery heretofore made, and if the conditions are favorable, we can expect with certitude that in the next few years wonders will be wrought by its application."

The Disturbing Influence of Solar Radiation on the Wireless Transmission of Energy

Electrical Review and Western Electrician, July 6, 1912

When Heinrich Hertz announced the results of his famous experiments in confirmation of the Maxwellian electromagnetic theory of light, the scientific mind at once leaped to the conclusion that the newly discovered dark rays might be used as a means for transmitting intelligible messages through space. It was an obvious inference, for heliography, or signaling by beams of light, was a well recognized wireless art. There was no departure in principle, but the actual demonstration of a cherished scientific idea surrounded the novel suggestion with a nimbus of originality and atmosphere of potent achievement. I also caught the fire of enthusiasm but was not long deceived in regard to the practical possibilities of this method of conveying intelligence.

Granted even that all difficulties were successfully overcome, the field of application was manifestly circumscribed. Heliographic signals had been flashed to a distance of 200 miles, but to produce Hertzian rays of such penetrating power as those of light appeared next to impossible, the frequencies obtainable through electrical discharges being necessarily of a much lower order. The rectilinear propagation would limit the action on the receiver to the extent of the horizon and entail interference of obstacles in a straight line joining the stations. The transmission would be subject to the caprices of the air and, chief of all drawbacks, the intensity of disturbances of this character would rapidly diminish with distance.

But a few tests with apparatus, far ahead of the art of that time, satisfied me that the solution lay in a different direction, and after a careful study of the problem I evolved a new plan which was fully described in my addresses before the Franklin Institute and National Electric Light Association in February and March, 1893. It was an extension of the transmission through a single wire without return, the practicability of which I had already demonstrated. If my ideas were rational, distance was of no consequence and energy could be conveyed from one to any point of the globe, and in any desired amount. The task was begun under the inspiration of these great possibilities.

While scientific investigation had laid bare all the essential facts relating to Hertz-wave telegraphy, little knowledge was available bearing on the system proposed by me. The very first requirement, of course, was the production of powerful electrical vibrations. To impart these to the earth in an efficient manner, to construct proper receiving apparatus, and develop other technical details could be confidently undertaken. But the all important question was, how would the planet be affected by the oscillations impressed upon it? Would not the capacity of the terrestrial system, composed of the earth and its conducting envelope, be too great? As to this, the theoretical prospect was for a long time discouraging. I found that currents of high frequency and potential, such as had to be necessarily employed for the purpose, passed freely through air moderately rarefied. Judging from these experiences, the dielectric stratum separating the two conducting spherical surfaces could be scarcely more than 20 kilometers thick and, consequently, the capacity would be over 220,000 microfarads, altogether too great to permit economic transmission of power to distances of commercial importance.

Another observation was that these currents cause considerable loss of energy in the air around the wire. That such waste might also occur in the earth's atmosphere was but a logical inference.

A number of years passed in efforts to improve the apparatus and to study the electrical phenomena produced. Finally my labors were rewarded and the truth was positively established; the globe did not act like a conductor of immense capacity and the loss of energy, due to absorption in the air, was insignificant. The exact mode of propagation of the currents from the source and the laws governing the electrical movement had still to be ascertained. Until this was accomplished the new art could not be placed on the plane of scientific engineering. One could bridge the greatest distance by sheer force, there being virtually no limit to the intensity of the vibrations developed by such a transmitter, but the installment of economic plants and the predetermination of the effects, as required in most practical applications, would be impossible.

Such was the state of things in 1899 when I discovered a new difficulty which I had never thought of before. It was an obstacle which could not be overcome by any improvement devised by man and of such nature as to fill me with apprehension that transmission of power without wires might never be quite practicable. I think it useful, in the present phase of development, to acquaint the profession with my investigations.

It is a well know fact that the action on a wireless receiver is appreciably weaker during the day than at night and this is attributed to the effect of sunlight on the elevated aerials, an explanation naturally suggested through an early observation of Heinrich Hertz. Another theory, ingenious but rather fine-spun, is that some of the energy of the waves is absorbed by ions or electrons, freed in sunlight and caused to move in the direction of propagation. *The Electrical Review and Western Electrician* of June 1, 1912, contains a report of a test, during the recent solar eclipse, between the station of the Royal Dock Yard in Copenhagen and the Blaavandshuk station on the coast of Jutland, in which it was demonstrated that the signals in that region became more distinct and reliable when the sunlight was partially cut off by the moon. The object of this communication is to show that in all the instances reported the weakening of the impulses was due to an entirely different cause.

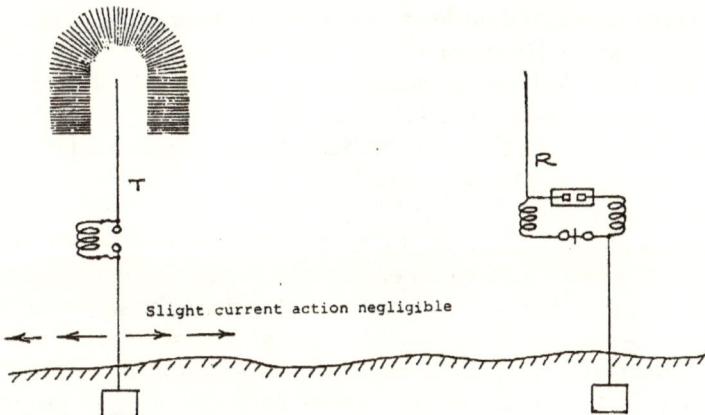

Fig. 1 — Hertz Wave System

It is indispensable to first dispel a few errors under which electricians have labored for years, owing to the tremendous momentum imparted to the scientific mind through the work of Hertz which has hampered independent thought and experiment. To facilitate understanding, attention is called to the annexed diagrams in which Fig. 1 and Fig. 2 represent, respectively, the well known arrangements of circuits in the Hertz-wave system and my own. In the former the transmitting and receiving conductors are separated from the ground through spark gaps, choking coils, and high remittances. This is necessary, as a ground connection greatly reduces the intensity of the radiation by cutting off half of the oscillator and also by increasing the length of the waves from 40 to 100 percent, according to the distribution of capacity and inductance. In the system devised by me a connection to earth, either directly or through a condenser is essential. The receiver, in the first case, is affected only by rays transmitted through the air, conduction being excluded; in the latter instance there is no appreciable radiation and the receiver is energized through the earth while an equivalent electrical displacement occurs in the atmosphere.

Now, an error which should be the focus of investigation for experts is, that in the arrangement shown in Fig. 1 the Hertzian effect has been gradually reduced through the lowering of frequency, so as to be negligible when the usual wavelengths are employed. That the energy is transmitted chiefly, if not wholly, by conduction can be demonstrated in a number of ways. One is to replace the vertical transmitting wire by a horizontal one of the same effective capacity, when it will be found that the action on the receiver is as before. Another evidence is afforded by quantitative measurement which proves that the energy received does not diminish with the square of the distance, as it should, since the Hertzian radiation propagates in a hemisphere. One more experiment in support of this view may be suggested. When transmission through the ground is prevented or impeded, as by severing the connection or otherwise, the receiver fails to respond, at least when the distance is considerable. The plain fact is that the Hertz waves emitted from the aerial are just as much of a loss of power as the short radiations of heat due to frictional waste in the wire. It has been contended that radiation and conduction might both be utilized in actuating the receiver, but this view is untenable in the light of my discovery of the wonderful law governing the movement of electricity through the globe, which may be conveniently expressed by the statement that the projection of the wave-lengths (measured along the surface) on the earth's diameter or axis of symmetry of movement are all equal. Since the surfaces of the zones so defined are the same the law can also be expressed by stating *that the current sweeps in equal times over equal terrestrial areas*. (See among others *"Handbook of Wireless Telegraph,"* by James Erskine-Murray.) Thus the velocity propagation through the superficial layers is variable, dependent on the distance from the transmitter, the mean value being n/2 times the velocity of light, while the ideal flow along the axis of propagation takes place with a speed of approximately 300,000 kilometers per second. To illustrate, the current from a transmitter situated at the Atlantic Coast will traverse that ocean—a distance of 4,800 kilometers—in less than 0.006 second with an average speed of 800,000 kilometers. If the signaling were done by Hertz waves the time required would be 0.016 second.

Bearing, then, in mind, that the receiver is operated simply by currents conducted along the earth as through a wire, energy radiated playing no part, it will be at once evident that the weakening of the impulses could not be due to any changes in the air, making it turbid or conductive, but should be traced to an effect interfering with the transmission of the current through the superficial layers of the globe. The solar radiations are the primary cause, that is true, not those of light, but of heat. The loss of energy, I have found, is due to the evaporation of the water on that side of the earth which is turned toward the sun, the conducting particles

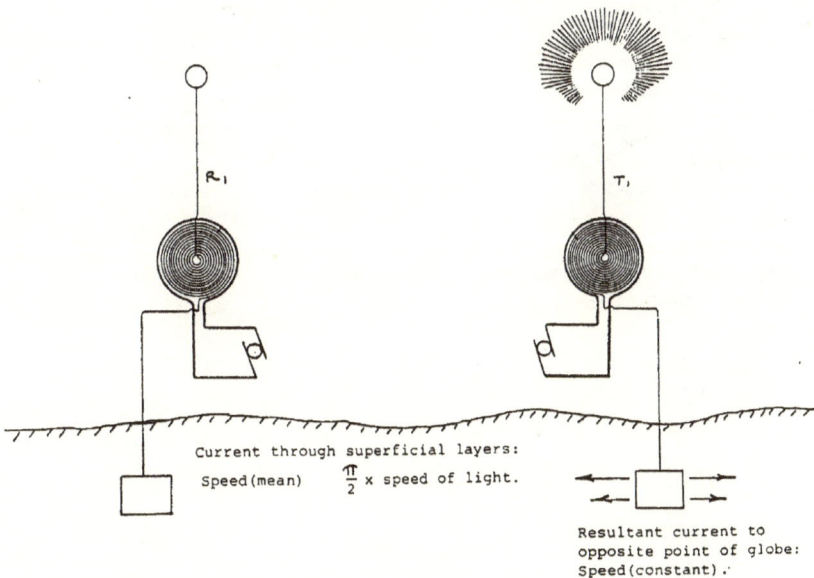

Current through superficial layers:
Speed(mean) $\frac{\pi}{2}$ x speed of light.

Resultant current to opposite point of globe: Speed(constant) .

Fig. 2 — System Devised by Tesla

carrying off more or less of the electrical charges imparted to the ground. This subject has been investigated by me for a number of years and on some future occasion I propose to dwell on it more extensively. At present it may be sufficient, for the guidance of experts, to state that the waste of energy is proportional to the product of the square of the electric density induced by the transmitter at the earth's surface and the frequency of the currents. Expressed in this manner it may not appear of very great practical significance. But remembering that the surface density increases with the frequency it may also be stated that the loss is proportional to the cube of the frequency. With waves 300 meters [1 MHZ] in length economic transmission of energy is out of the question, the loss being too great. When using wave-lengths of 6,000 meters [50 kHz] it is still noticeable though not a serious drawback. With wave-lengths of 12,000 meters [25 kHz] it becomes quite insignificant and on this fortunate fact rests the future of wireless transmission of energy.

To assist investigation of this interesting and important subject, Fig. 3 has been added, showing the earth in the position of summer solstice with the transmitter just emerging from the shadow. Observation will bring out the fact that the weakening is not noticeable until the aerials have reached a position, with reference to the sun, in which the evaporation of the water is distinctly more rapid. The maximum will not be exactly when the angle of incidence of the suns rays is greatest, but some time after. It is noteworthy that the experimenters who watched the effect of the recent eclipse, above referred to, have observed the delay.

Fig. 3 — Illustrating Disturbing Effect of the Sun on Wireless Transmission.

Nikola Tesla's Plan to Keep "Wireless Thumb" on Ships At Sea
New York Press — Nov. 9

Suggests Transmitters Powerful Enough to Cause the Earth to Vibrate at the Poles and Equator.

He would Determine Vessel's Latitude and Longitude by Measuring the Length of Electric Waves.

Nikola Tesla has come forward to refute the claims of men who recently excited the scientific world with announcements of discovery and invention calculated to crowd the bugbear of scientific warfare back into the primer class, and to safeguard the lives of seafarers. First he takes up and disposes of the announcement of an invention said to enable a receiving ship equipped with wireless to tell the longitude and latitude of a sending ship without the latter vessel offering its own calculations. "It hasn't been done, and it probably will be years before the means for so doing can be applied successfully," he says. As to the power of ultra-violet rays to explode the powder magazine of a warship from a distance, he insists it can't be done through that medium. If charges of powder have been so exploded, he contends, the detonation was accomplished with the familiar waves now utilized by the wireless. But Mr. Tesla admits that in all probability there will come a time when science has so harnessed and developed the means at hand that such results may be obtained. Mr. Tesla sets forth for readers of The Press his views on the two subjects as follows:

The first and incomplete announcements of technical advances should always be taken with a grain of salt. It is true that the newspapers are getting more and more accurate and reliable in putting forth such information, but, nevertheless, the news frequently is misleading.

For instance, not long ago reports widely circulated that powder had been exploded at distance by infra-red or ultra-violet rays, and that a British battleship had been used in a test of this kind, which proved successful. The dispatches gave great opportunity to sensational speculation, but the truth is that there was no novelty whatever in what was done.

A mine or magazine 'may have been blown up, but this was accomplished in a well-known manner through the application of a kind of electrical waves which are now generally adopted in the transmission of signals without wires. Similar experiments were performed in this country many years ago by myself and others, and quite recently John Hays Hammond, Jr., has done credible work in this direction through the application of an art which has been named "Telautomatics," or wireless control of moving mechanism at a distance.

By means of such telautomatic vessels, surface, and submarine or aerial, a perfect system of coast defense can be established. Torpedoes on this plan also can be controlled from battleships, and there is no doubt they sooner or later will be adopted and their introduction will have a revolutionary effect on the methods of warfare.

The results described are, however, not impossible. It is quite practicable to explode by rays of light a mine at a distance, as by acting, on a mixture of chlorine and hydrogen. Certain dark rays also can be employed to produce destructive effects. As far back as 1897, I disclosed before the New York Academy of Sciences the discovery that Roentgen, or X-rays, projected from certain bulbs have the property of strongly charging an electrical condenser at a distance. The energy so accumulated readily can be discharged and cause the ignition of some explosive compound.

But ultra-violet rays are of very short wave lengths and cannot penetrate steel shells, while the longer and more penetrative waves of the infra-red rays are chemically much less active. There is no doubt in my mind that we soon shall be able to project energy at a distance not only in small, but in large amounts, and what the effect of such an achievement will be on existing conditions, words cannot express.

As regards the determination of latitude and longitude of a vessel at sea by wireless, there is nothing in use as yet which would make such direct observation possible. Some suggestions, however, which I have since many years advocated, have been adopted. They are the flashing of time signals over a wide area and the employment of an instrument known as a wireless compass.

These means enable an expert on a vessel to ascertain the exact hour at any sending station within reach, and also, in an imperfect manner, the direction in which it is situated, and from these data it is possible to get a rough idea of the position of the ship relative to the points of reference.

A perfect means for determining not only such and other data important to the navigator already is available, but it may require years to apply it. I refer to the use of the stationary waves, which were discovered by me fourteen years ago. The subject is too technical to be explained in detail, but the average reader can be made to understand the general principle.

The earth is a conductor of electricity, and as such has its own electrical period of vibration. The time of one complete swing is about one-twelfth of a second. In other words, this is the interval the current requires in passing to, and returning from, the diametrically opposite point of the globe.

Now, the wonderful fact is, that notwithstanding its immense size, the earth responds to a great number of vibrations and can be resonantly excited just like a wire of limited dimensions. When this takes place there are formed on its surface stationary parallel circles of equal electrical activity, which can be revealed by properly attuned instruments.

Imagine that a transmitter capable of exciting the earth were placed at one of the Poles. Then the crests and hollows of the stationary waves would be in parallel circles with their planes at right angles to the axis of the earth, and from readings of a properly graduated instrument the distance of a vessel carrying the same from the Pole could be at once read, giving accurately the geographical latitude.

In like manner, if a transmitter were placed at a point on the Equator, the longitude could be precisely determined by the same means. But the best plan would be to place three transmitters at properly chosen points on the globe so as to establish three non-interferable systems of stationary waves at right angles to one another. If this were done, innumerable results of the greatest practical value could be realized.

Nikola Tesla Sees A Wireless Vision

New York Times, Oct. 3, 1915

Nikola Tesla announced to The Times last night that he had received a patent on an invention which would not only eliminate static interference, the present bugaboo of wireless telephony, but would enable thousands of persons to talk at once between wireless stations and make it possible for those talking to see one another by wireless, regardless of the distance separating them. He said also that with his wireless station now in the process of construction on Long Island he hoped to make New York one of the central exchanges in a world system of wireless telephony.

Mr. Tesla has been working on wireless problems for many years. Yesterday he exhibited an article published in the Electrical World eleven years ago, in which he predicted not only wireless telephony on a commercial basis but that it would be possible to identify the voice of an acquaintance over any distance. That its operator in Hawaii was able to distinguish the voice of an engineer friend at Arlington, Va., was announced by the American Telephone and Telegraph Company as the most marked triumph of its communication by wireless telephone from the naval radio station at Arlington to Pearl Harbor, Hawaii, a distance of 4,000 miles.

The inventor, who has won fame by his electrical inventions, dictated this statement yesterday.

"The experts carrying out this brilliant experiment are naturally deserving of great credit for the skill they have shown in perfecting the devices. These are of two kinds: First, those serving to control transmission, and, second, those magnifying the received impulse. That the control of transmission is perfect is plain to experts from the fact the Arlington, Mare Island, and Pearl Harbor plants are all inefficient and that the distance of telephonic transmission is equal to that of telegraphic transmission. It is also perfectly apparent that the chief merit of the application lies in the magnification of the microphonic impulse. It must not be imagined that we deal here with new discoveries. The improvement simply concerns the control of the transmitted and the magnification of the received impulse, but the wireless system is the same. This can never be changed.

"That it is practicable to project the human voice not only to a distance of 5,000 miles, but clear across the globe, I demonstrated by experiments in Colorado in 1899. Among my publications I would refer to an article in the Electrical World of March 5, 1904, but describing really tests I made in 1899. The facts which I pointed out in the article were of much greater significance than that of the experiments reported, although this should be taken in a scientific sense, as the experiments were simply scientific demonstrations. I pointed out then that the modulations of the human voice can be reproduced more clearly through the earth than through wire. It is difficult for the layman to understand, but it is an absolute fact that transmission through the earth with the proper apparatus is not more difficult than the sending of a message on a wire strung across a room. This wonderful property of the planet, that, electrically speaking, is through its very bigness, small, is of incalculable significance for the future of mankind.

"These tests made between Washington and Honolulu will act as an immense stimulus to wireless telephony and would be of much more value to the world if the principles of the transmission were understood. But they are not. Even now, fifteen years after the fundamental principles have been demonstrated and the possibilities shown, there are many experts in the dark.

"For instance, it is claimed that static disturbances will fatally interfere with the transmission, while, as a matter of fact, there is no static disturbance possible in properly designed transmission and receiving circuits. Quite recently I have described in a patent, circuits which are absolutely immune to static and other interferences - so much so that when a telephone is attached, there is absolute silence, even lightning in the immediate vicinity not producing a click of the diaphragm, while in the ordinary telephonic conversation there are all kinds of noises. Transmission without static interference has many wonderful properties, besides, first of which is that unlimited amounts of power can be transmitted with very small loss.

"Another contention is that there can be no secrecy in wireless telephone conversation. I say it is absurd to raise this contention when it is positively demonstrated by experiments that the earth is more suitable for transmission than any wire could ever be. A. wireless telephone conversation can be made as secret as thought.

"I have myself erected a plant for the purpose of connecting by wireless telephone the chief centers of the world, and from this plant as many as a hundred will be able to talk absolutely without interference and with absolute secrecy. This plant would simply be connected with the telephone central exchange of New York City, and any subscriber will be able to talk to any other telephone subscriber in the world, and all this without any change in his apparatus. This plan has been called my "world system." By the same means I propose also to transmit pictures and project images, so that the subscriber will not only hear the voice, but see the person to whom he is talking. Pictures transmitted over wires is a perfectly simple art practiced today. Many inventors have labored on it, but the chief credit is due Professor Korn of Munich.

"His apparatus can be attached to a wireless plant and at any other wireless plant can be reproduced. I have undertaken this in the hope of establishing a service which would greatly facilitate the work of the press. A picture could be sent from a battlefield in Europe to New York in five minutes if the proper instruments were available.

"A further advantage would be that transmission is instant and free of the unavoidable delay experienced with the use of wires and cables. As I have already made known, the current passes through the earth, starting from the transmission station with infinite speed, slowing down to the speed of light at a distance of 6,000 miles, then increasing in speed from that region and reaching the receiving station again with infinite velocity.

"It's all a wonderful thing. Wireless is coming to mankind in its full meaning like a hurricane some of these days. Some day there will be, say, six great wireless telephone stations in a world system connecting all the inhabitants on this earth to one another, not only by voice, but by sight. It's surely coming.

The Effect of Static on Wireless Transmission

Electoral Experimenter—January, 1919

A few statements regarding these phenomena in response to a request of the *Electrical Experimenter* may be useful at the present time in view of the increasing and importance of the
subject.

The commercial application of the art has led to the construction of larger transmitters and multiplication of their number, greater distances had to be covered and it became imperative to employ receiving devices of ever increasing sensitiveness. All these and other changes have cooperated in emphasizing the trouble and seriously impairing the reliability and value of the plans. To such a degree has this been the ease that conservative business men and financiers have come to look upon this method of conveying intelligence as one offering but very limited possibilities, and the Government has deemed it advisable to assume control. This unfortunate state of affairs, fatal to enlistment of capital and healthful competitive development, could have been avoided had electricians not remained to this day under the spell of a delusive theory and had the practical exploiters of this advance not permitted enterprise to outrun technical competence.

With the publication of Dr . Heinrich Hertz's classical researches it was an obvious inference that the dark rays investigated by him could be used for signaling purposes, as those of light in heliography, and the first steps in this direction were made with his apparatus which, in 1896, was found capable of actuating receivers at a distance of a few miles. Three years prior to this, however, in lectures before the Franklin Institute and National Electric Light Association, I had described a wireless system radically opposite to the Hertzian in principle inasmuch as it depended on currents conducted through the earth instead of on radiations propagated through the atmosphere, presumably in straight lines.

The apparatus then outlined by me consisted of a transmitter comprising a primary circuit excited from an alternator or equivalent source of electrical energy and a high potential secondary resonant circuit, connected with its terminals to ground and to an elevated capacity, and a similar tuned receiving circuit including the operative device. On that occasion I expressed myself confidently on the feasibility of flashing in this manner not only signals to any terrestrial distance but Transmitting power in unlimited amounts for all sorts of industrial purposes. The discoveries made and experimental results attained I made with a wireless power-plant erected in 1899 some of which were disclosed in the *Century Magazine* of June, 1900, and several U. S. patents subsequently granted to me have, I believe, borne-out strikingly my foresight. In the meantime the Hertzian arrangements were gradually modified, one feature after another being abandoned, so that now not a vestige of them can be found and my system of four tuned circuits has been universally adopted, not only in its fundamentals but in every detail as the "quenched sparks," "ticker," "tone wheel," high frequency and rotating field alternators, forms of discharges and mercury breaks, frequency changers, coils, condensers, regulating methods, and devices, etc. This fact would give me supreme satisfaction were it not that the engineers, misinterpreting the nature of the effects, are making installments so defective in construction and mode of operation as to preclude the possibility of the great realization which might be brought within easy reach by proper application of the underlying principles and one of which—the most desirable at present—is the complete elimination of all static and other interference.

During the past few years several emphatic announcements have been made that a perfect solution of this problem had been discovered, but it was manifest from a casual perusal of these publications that the experts were ignoring certain truths of vital bearing on the question, and so long as this was the case no such claim could be substantiated. I achieved early success by keeping these steadily in mind and applying my efforts from the outset in the right and correct scientific direction.

I may contribute to the clearness of the subject in answering a question which I have been asked by the Editors of the *Electrical Experimenter* with reference to the report contained in the last issue, that signals had been received around the globe. An achievement the practicability of which I have fully demonstrated by experiment eighteen yeas ago.

The question is, how can Hertz waves be conveyed to such a distance in view of the curvature of the earth? A few words will be sufficient to show the absurdity of the prevailing opinion propounded in text books.

We are living on a conducting globe surrounded by a thin layer of insulating air, above which is a rarefied and conducting atmosphere. If the earth is represented by a sphere of 12 ½ inch radius, then the layer which may be considered insulating for high frequency currents of great tension is less than 1/64 of an inch thick. It is odd that the Hertz waves, emanating from a transmitter, get to the distant receiver by successive reflections. The utter impossibility of this will be evident when it is shown by a simple calculation that the amount of energy received , even if it could be collected in its totality, is infinitesimal and would not actuate the most sensitive instrument known were it magnified many million times. The fact is these waves have no perceptible influence on a receiver if situated at a much smaller distance. It should be remembered, moreover, that since the first attempts the wave lengths have been increased until those advocated by me were adopted, in which this form of radiation has been reduced to one-billionth.

When a circuit, connected to ground and to an elevated capacity oscillates two effects separate and distinct arc produced; Hertz waves are radiated in a direction at right angles to the axis of symmetry of the conductor, *and simultaneously a current is passed through the earth*. The former propagates with the speed of light, the latter with a velocity proportionate to the cosecant of an angle which from the origin to the opposite point of the globe varies from zero to 180°. Expressed in words, at the start the speed is infinite and diminishes, first rapidly and then slowly until a quadrant is traversed when the current proceeds with the speed of light. From that region on the velocity gradually increases becoming infinite at the opposite point of the globe. In a patent granted to me in April, 1905, I have summed up this law of propagation in the statement that the projections of all half waves on the axis of symmetry of movement are equal, which means that the successive half waves, though of different length, cover exactly the same area. In the near future many wonderful results will be obtained by taking advantage of this fact.

Tesla's Static Eliminator, Patented and Used by Him over Twenty Years Ago. It Will Be Fully Described in an Early Issue of the Electrical Experimenter

There is a vast difference between these two forms of wave movement in their bearing on the transmission. The Hertz waves represent energy which is radiated and unrecoverable. The current energy on the other hand, is preserved and can be recovered theoretically at least, in its entirety. If the experts will free themselves from that illusions under which they are laboring they will find that to overcome static disturbances all that is needed is the properly constructed transmitter and receiver without any additional devices or preventives. I have, however, devised several forms of apparatus eliminating statics even in the present defective wireless installations in which they are magnified many times. Such a form of instrument which I have used successfully is shown in the annexed photograph. These phenomena have been studied by me for a number of years and I have found that there are nine or ten different causes that tend to intensify them, and in due course I shall give a full description of the various improvements I have made, in the *Electrical Experimenter*. For the present I would only point out that in order to perfectly eliminate the static interference, it is indispensable to redesign the whole wireless apparatus as now employed. The sooner this is understood the better it will be for the further evolution of the Art.

The True Wireless
Electrical Experimenter, May 1919

In this remarkable and complete story of his discovery of the "True Wireless" and the principles upon which transmission and reception, even in the present day systems, are based, Dr. Nikola Tesla shows us that he is indeed the "Father of the Wireless." To him the Hertz wave theory is a delusion; it looks sound from certain angles, but the facts tend to prove that it is hollow and empty. He convinces us that the real Hertz waves are blotted out after they have traveled but a short distance from the sender. It follows, therefore, that the measured antenna current is no indication of the effect, because only a small part of it is effective at a distance. The limited activity of pure Hertz wave transmission and reception is here clearly explained, besides showing definitely that in spite of themselves, the radio engineers of today are employing the original Tesla tuned oscillatory system. He shows by examples with different forms of aerials that the signals picked up by the instruments must actually be induced by earth currents—not etheric space waves. Tesla also disproves the "Heaviside layer" theory from his personal observations and tests. —Editor

Ever since the announcement of Maxwell's electro-magnetic theory scientific investigators all the world over had been bent on its experimental verification. They were convinced that it would be done and lived in an atmosphere of eager expectancy, unusually favorable to the reception of any evidence to this end. No wonder then that the publication of Dr. Heinrich Hertz's results caused a thrill as had scarcely ever been experienced before. At that time I was in the midst of pressing work in connection with the commercial introduction of my system of power transmission, but, nevertheless, caught the fire of enthusiasm and fairly burned with desire to behold the miracle with my own eyes. Accordingly, as soon as I had freed myself of these imperative duties and resumed research work in my laboratory on Grand Street, New York, I began, parallel with high frequency alternators, the construction of several forms of apparatus with the object of exploring the field opened up by Dr. Hertz. Recognizing the limitations of the devices he had employed, I concentrated my attention on the production of a powerful induction coil but made no notable progress until a happy inspiration led me to the invention of the oscillation transformer. In the latter part of 1891 I was already so far advanced in the development of this new principle that I had at my disposal means vastly superior to those of the German physicist. All my previous efforts with Rhumkorf coils had left me unconvinced, and in order to settle my doubts I went over the whole ground once more, very carefully, with these improved appliances. Similar phenomena were noted, greatly magnified in intensity, but they were susceptible of a different and more plausible explanation. I considered this so important that in 1892 I went to Bonn, Germany, to confer with Dr. Hertz in regard to my observations. He seemed disappointed to such a degree that I regretted my trip and parted from him sorrowfully. During the succeeding years I made numerous experiments with the same object, but the results were uniformly negative. In 1900, however, after I had evolved a wireless transmitter which enabled me to obtain electro-magnetic activities of many millions of horse-power, I made a last desperate attempt to prove that the disturbances emanating from the oscillator were ether vibrations akin to those of light, but met again with utter failure. For more than eighteen years I have been reading treatises, reports of scientific transactions, and articles on Hertz-wave telegraphy, to keep myself informed, but they have always impressed me like works of fiction.

The history of science shows that theories are perishable. With every new truth that is revealed we get a better understanding of Nature and our conceptions and views are modified. Dr. Hertz did not discover a new principle. He merely gave material support to hypothesis which had been long ago formulated. It was a perfectly well-established fact that a circuit, traversed by a periodic current, emitted some kind of space waves, but we were in ignorance as to their character. He apparently gave an experimental proof that they were transversal vibrations in the ether. Most people look upon this as his great accomplishment. To my mind it seems that his immortal merit was not so much in this as in the focusing of the investigators' attention on the processes taking place in the ambient medium. The Hertz-wave theory, by its fascinating hold on the imagination, has stifled creative effort in the wireless art and retarded it for twenty-five years. But, on the other hand, it is impossible to over-estimate the beneficial effects of the powerful stimulus it has given in many directions.

As regards signaling without wires, the application of these radiations for the purpose was quite obvious. When Dr. Hertz was asked whether such a system would be of practical value, he did not think so, and he was correct in his forecast. The best that might have been expected was a method of communication similar to the heliographic and subject to the same or even greater limitations.

In the spring of 1891 I gave my demonstrations with a high frequency machine before the American Institute of Electrical Engineers at Columbia College, which laid the foundation to a new and far more promising departure. Altho the laws of electrical resonance were well known at that time and my lamented friend, Dr. John Hopkinson, had even indicated their specific application to an alternator in the Proceedings of the Institute of Electrical Engineers, London, Nov.13, 1889, nothing had been done towards the practical use of this knowledge and it is probable that those experiments of mine were the first public exhibition with resonant circuits, more particularly of high frequency. While the spontaneous success of my lecture was due to spectacular features, its chief import was in showing that all kinds of devices could be operated thru a single wire without return. This was the initial step in the evolution of my wireless system. The idea presented itself to me that it might be possible, under observance of proper conditions of resonance, to transmit electric energy thru the earth, thus dispensing with all artificial conductors. Anyone who might wish to examine impartially the merit of that early suggestion must not view it in the light of present day science. I only need to say that as late as 1893, when I had prepared an elaborate chapter on my wireless system, dwelling on its various instrumentalities and future prospects, Mr. Joseph Wetzler and other friends of mine emphatically protested against its publication on the ground that such idle and far-fetched speculations would injure me in the opinion of conservative business men. So it came that only a small part of what I had intended to say was embodied in my address of that year before the Franklin Institute and National Electric Light Association under the chapter "On Electrical Resonance." This little salvage from the wreck has earned me the title of "Father of the Wireless" from many well-disposed fellow workers, rather than the invention of scores of appliances which have brought wireless transmission within the reach of every young amateur and which, in a time not distant, will lead to undertakings overshadowing in magnitude and importance all past achievements of the engineer.

The popular impression is that my wireless work was begun in 1893, but as a matter of fact I spent the two preceding years in investigations, employing forms of apparatus, some of which were almost like those of today. It was clear to me from the very start that the successful consummation could only be brought about by a number of radical improvements. Suitable high frequency generators and electrical oscillators had first to be produced. The energy of these had to be transformed in effective transmitters and collected at a distance in proper receivers. Such a system would be manifestly circumscribed in its usefulness if all extraneous interference were not prevented and exclusiveness secured. In time, however, I recognized that devices of this kind, to be most effective and efficient, should be designed with due regard to the physical properties of this planet and the electrical conditions obtaining on the same. I will briefly touch upon the salient advances as they were made in the gradual development of the system.

The high frequency alternator employed in my first demonstrations is illustrated in Fig. 1. It comprised a field ring, with 384 pole projections and a disc armature with coils wound in one single layer which were connected in various ways according to requirements. It was an excellent machine for experimental purposes, furnishing sinusoidal currents of from 10,000 to 20,000 cycles per second. The output was comparatively large, due to the fact that as much as 30 amperes per square millimeter could be past thru the coils without injury.

Fig. 1. Alternator of 10.000 Cycles p.s., Capacity 10K.W.. Which Was Employed by Tesla in His First Demonstrations of High Frequency Phenomena Before the American Institute of Electrical Engineers at Columbia College, May 20, 1891.

The diagram in Fig. 2 shows the circuit arrangements as used in my lecture. Resonant conditions were maintained by means of a condenser subdivided into small sections, the finer adjustments being effected by a movable iron core within an inductance coil. Loosely linked with the latter was a high tension secondary which was tuned to the primary.

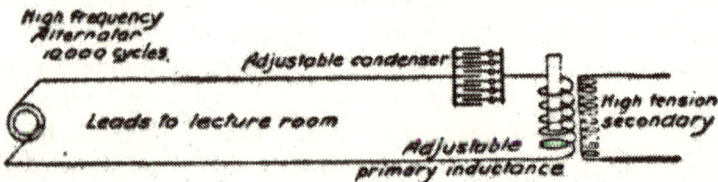

Fig. 2. Diagram Illustrating the Circuit Connections and Tuning Devices Employed by Tesla In His Experimental Demonstrations Before the American Institute of Electrical Engineers With the High Frequency Alternator Shown in Fig. 1.

The operation of devices thru a single wire without return was puzzling at first because of its novelty, but can be readily explained by suitable analogs. For this purpose reference is made to Figs. 3 and 4.

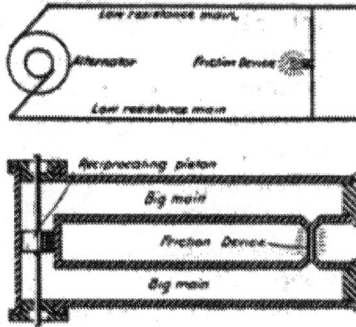

Fig. 3. Electric Transmission Thru Two Wires and Hydraulic Analog.

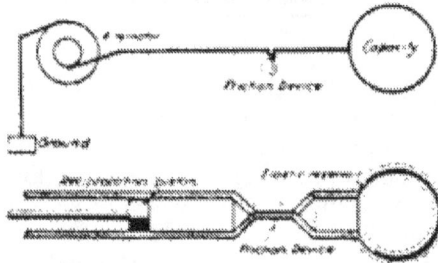

Fig. 4. Electric Transmission Thru a Single Wire Hydraulic Analog.

In the former the low resistance electrical conductors are represented by pipes of large cross section, the alternator by an oscillating piston and the filament of an incandescent lamp by a minute channel connecting the pipes. It will be clear from a glance at the diagram that very slight excursions of the piston would cause the fluid to rush with high velocity thru the small channel and that virtually all the energy of movement would be transformed into heat by friction, similarly to that of the electric current in the lamp filament.

The second diagram will now be self-explanatory. Corresponding to the terminal capacity of the electric system an elastic reservoir is employed which dispenses with the necessity of a return pipe. As the piston oscillates the bag expands and contracts, and the fluid is made to surge thru the restricted passage with great speed, this resulting in the generation of heat as in the incandescent lamp. Theoretically considered, the efficiency of conversion of energy should be the same in both cases.

Granted, then, that an economic system of power transmission thru a single wire is practicable, the question arises how to collect the energy in the receivers. With this object attention is called to Fig. 5, in which a conductor is shown excited by an oscillator joined to it at one end. Evidently, as the periodic impulses pass thru the wire, differences of potential will be created along the same as well as at right angles to it in the surrounding medium and either of these may be usefully applied. Thus at a, a circuit comprising an inductance and capacity is resonantly excited in the transverse, and at b, in the longitudinal sense. At c, energy is collected in a circuit parallel to the conductor but not in contact with it, and again at d, in a circuit which is partly sunk into the conductor and may be, or not, electrically connected to the same. It is important to keep these typical dispositions in mind, for however the distant actions of the oscillator might be modified thru the immense extent of the globe the principles involved are the same.

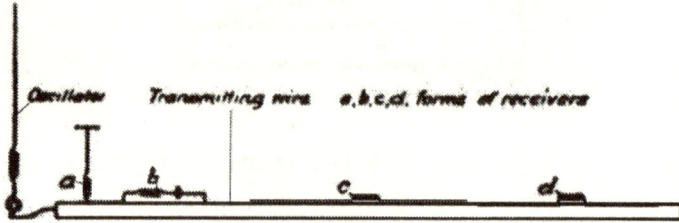

Fig. 5. Illustrating Typical Arrangements for Collecting Energy In a System of Transmission Thru a Single Wire.

Consider now the effect of such a conductor of vast dimensions on a circuit exciting it. The upper diagram of Fig. 6 illustrates a familiar oscillating system comprising a straight rod of self-inductance 2L with small terminal capacities cc and a node in the center. In the lower diagram of the figure a large capacity C is attached to the rod at one end with the result of shifting the node to the right, thru a distance corresponding to self-inductance X. As both parts of the system on either side of the node vibrate at the same rate, we have evidently, (L+X)c = (L-X)C from which X = L(C-c/C+c). When the capacity C becomes commensurate to that of the earth, X approximates L, in other words, the node is close to the ground connection. The exact determination of its position is very important in the calculation of certain terrestrial electrical and geodetic data and I have devised special means with this purpose in view.

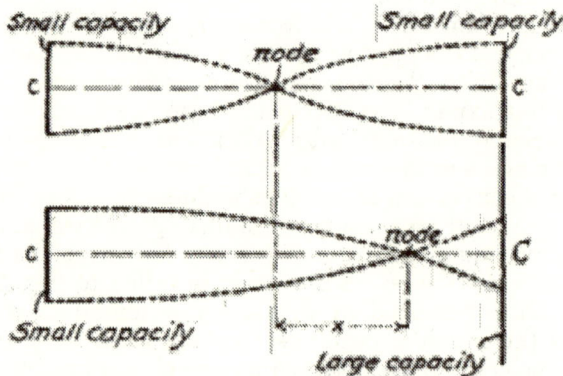

Fig. 6. Diagram Elucidating Effect of Large Capacity on One End.

My original plan of transmitting energy without wires is shown in the upper diagram of Fig. 7, while the lower one illustrates its mechanical analog, first published in my article in the Century Magazine of June, 1900. An alternator, preferably of high tension, has one of its terminals connected to the ground and the other to an elevated capacity and impresses its oscillations upon the earth. At a distant point a receiving circuit, likewise connected to ground and to an elevated capacity, collects some of the energy and actuates a suitable device. I suggested a multiplication of such units in order to intensify the effects, an idea which may yet prove valuable. In the analog two tuning forks are provided, one at the sending and the other at the receiving station, each having attached to its lower prong a piston fitting in a cylinder. The two cylinders communicate with a large elastic reservoir filled with an incompressible fluid. The vibrations transmitted to either of the tuning forks excite them by resonance and, thru electrical contacts or otherwise, bring about the desired result. This, I may say, was not a mere mechanical illustration, but a simple representation of my apparatus for submarine signaling, perfected by me in 1892, but not appreciated at that time, altho more efficient than the instruments now in use.

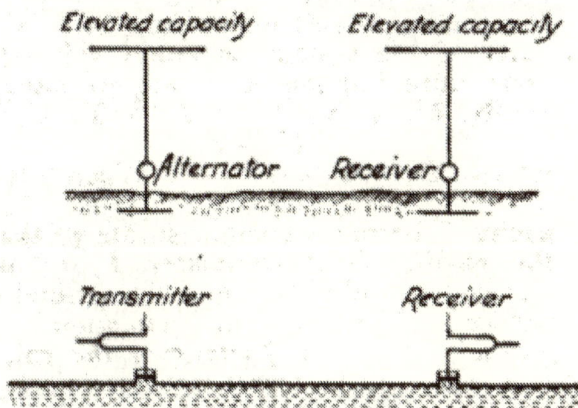

Fig. 7. Transmission of Electrical Energy Thru the Earth as Illustrated in Tesla's Lectures Before the Franklin Institute and Electric Light Association in February and March, 1893, and Mechanical Analog of the Same.

The electric diagram in Fig. 7, which was reproduced from my lecture, was meant only for the exposition of the principle. The arrangement, as I described it in detail, is shown in Fig. 8. In this case an alternator energizes the primary of a transformer, the high tension secondary of which is connected to the ground and an elevated capacity and tuned to the impressed oscillations. The receiving circuit consists of an inductance connected to the ground and to an elevated terminal without break and is resonantly responsive to the transmitted oscillations. A specific form of receiving device was not mentioned, but I had in mind to transform the received currents and thus make their volume and tension suitable for any purpose. This, in substance, is the system of today and I am not aware of a singe authenticated instance of successful transmission at considerable distance by different instrumentalities. It might, perhaps, not be clear to those who have perused my first description of these improvements that, besides making known new and efficient types of apparatus, I gave to the world a wireless system of potentialities far beyond anything before conceived. I made explicit and repeated statements that I contemplated transmission, absolutely unlimited as to terrestrial distance and amount of energy. But, altho I have overcome all obstacles which seemed in the beginning unsurmountable and found elegant solutions of all the problems which confronted me, yet, even at this very day, the majority of experts are still blind to the possibilities which are within easy attainment.

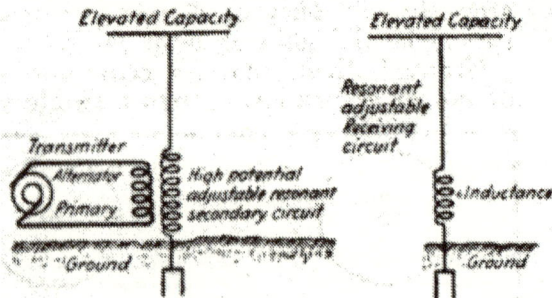

F i g .   8 .   T e s l a ' s
System of Wireless Transmission Thru the Earth as Actually Exposed In His Lectures Before the Franklin Institute and Electric Light Association in February and March, 1893.

My confidence that a signal could be easily flashed around the globe was strengthened thru the discovery of the "rotating brush," a wonderful phenomenon which I have fully described in my address before the Institution of Electrical Engineers, London, in 1892 [Experiments with Alternate Currents of High Potential and High Frequency], and which is illustrated in Fig. 9. This is undoubtedly the most delicate wireless detector known, but for a long time it was hard to produce and to maintain in the sensitive state. These difficulties do not exist now and I am looking to valuable applications of this device, particularly in connection with the high-speed photographic method, which I suggested, in wireless, as well as in wire, transmission.

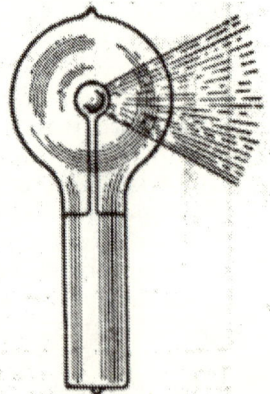

Fig. 9. The Forerunner of the Audion-
t he Most Sensitive Wireless Detector
Known, as Described by Tesla In His
Lecture Before the Institution of Electrical Engineers, London, February, 1892.

Possibly the most important advances during the following three
or four years were my system of concatenated tuned circuits and
methods of regulation, now universally adopted. The intimate
bearing of these inventions on the development of the wireless art
will appear from Fig. 10, which illustrates an arrangement
described in my U.S. Patent No. 568,178 of September 22, 1896, and
corresponding dispositions of wireless apparatus. The captions of
the individual diagrams are thought sufficiently explicit to
dispense with further comment. I will merely remark that in this
early record, in addition to indicating how any number of resonant
circuits may be linked and regulated, I have shown the advantage
of the proper timing of primary impulses and use of harmonics. In
a farcical wireless suit in London, some engineers, reckless of their
reputation, have claimed that my circuits were not at all attuned;
in fact they asserted that I had looked upon resonance as a sort of
wild and untamable beast!

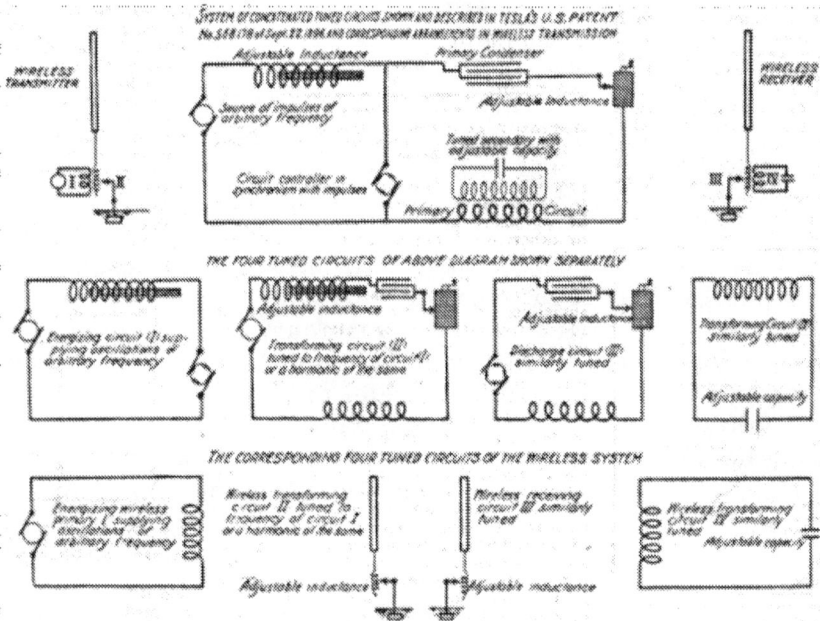

Fig. 10. Tesla's System of Concatenated Tuned Circuits Shown and
Described In U. S. Patent No. 568,178 of September 22, 1896, and
Corresponding Arrangements in Wireless Transmission.

It will be of interest to compare my system as first described in a Belgian patent of 1897 with the Hertz-wave system of that period. The significant differences between them will be observed at a glance. The first enables us to transmit economically energy to any distance and is of inestimable value; the latter is capable of a radius of only a few miles and is worthless. In the first there are no spark-gaps and the actions are enormously magnified by resonance. In both transmitter and receiver the currents are transformed and rendered more effective and suitable for the operation of any desired device. Properly constructed, my system is safe against static and other interference and the amount of energy which may be transmitted is billions of times greater than with the Hertzian which has none of these virtues, has never been used successfully and of which no trace can be found at present.

A well-advertised expert gave out a statement in 1899 that my apparatus did not work and that it would take 200 years before a message would be flashed across the Atlantic and even accepted stolidly my congratulations on a supposed great feat. But subsequent examination of the records showed that my devices were secretly used all the time and ever since I learned of this I have treated these Borgia-Medici methods with the contempt in which they are held by all fair-minded men. The wholesale appropriation of my inventions was, however, not always without a diverting side. As an example to the point I may mention my oscillation transformer operating with an air gap. This was in turn replaced by a carbon arc, quenched gap, an atmosphere of hydrogen, argon or helium, by a mechanical break with oppositely rotating members, a mercury interrupter or some kind of a vacuum bulb and by such tours de force as many new "systems" have been produced. I refer to this of course, without the slightest ill-feeling, let us advance by all means. But I cannot help thinking how much better it would have been if the ingenious men, who have originated these "systems," had invented something of their own instead of depending on me altogether.

Before 1900 two most valuable improvements were made. One of these was my individualized system with transmitters emitting a wave-complex and receivers comprising separate tuned elements cooperatively associated. The underlying principle can be explained in a few words. Suppose that there are n simple vibrations suitable for use in wireless transmission, the probability that any one tune will be struck by an extraneous disturbance is $1/n$. There will then remain $n-1$ vibrations and the chance that one of these will be excited is 1/n-1 hence the probability that two tunes would be struck at the same time is 1/n(n-1). Similarly, for a combination of three the chance will be 1/n(n-1)(n-2) and so on. It will be readily seen that in this manner any desired degree of safety against the statics or other kind of disturbance can be attained provided the receiving apparatus is so designed that is operation is possible only thru the joint action of all the tuned elements. This was a difficult problem which I have successfully solved so that now any desired number of simultaneous messages is practicable in the transmission thru the earth as well as thru artificial conductors.

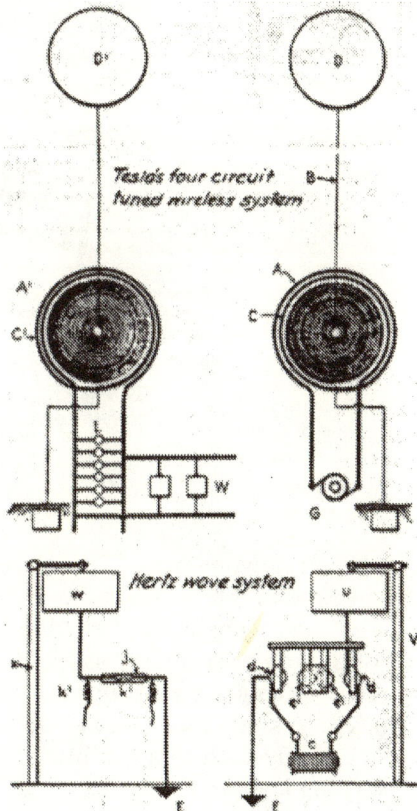

Fig. 11. Tesla's Four Circuit Tuned System Contrasted With the Contemporaneous Hertz-wave System.

The other invention, of still greater importance, is a peculiar oscillator enabling the transmission of energy without wires in any quantity that may ever be required for industrial use, to any distance, and with very high economy. It was the outcome of years of systematic study and investigation and wonders will be achieved by its means.

The prevailing misconception of the mechanism involved in the wireless transmission has been responsible for various unwarranted announcements which have misled the public and worked harm. By keeping steadily in mind that the transmission thru the earth is in every respect identical to that thru a straight wire, one will gain a clear understanding of the phenomena and will be able to judge correctly the merits of a new scheme. Without wishing to detract from the value of any plan that has been put forward I may say that they are devoid of novelty. So for instance in Fig. 12 arrangements of transmitting and receiving circuits are illustrated, which I have described in my U.S. Patent No. 613,809 of November 8, 1898 on a Method of and Apparatus for Controlling Mechanism of Moving Vessels or Vehicles, and which have been recently dished up as original discoveries. In other patents and technical publications I have suggested conductors in the ground as one of the obvious modifications indicated in Fig. 5.

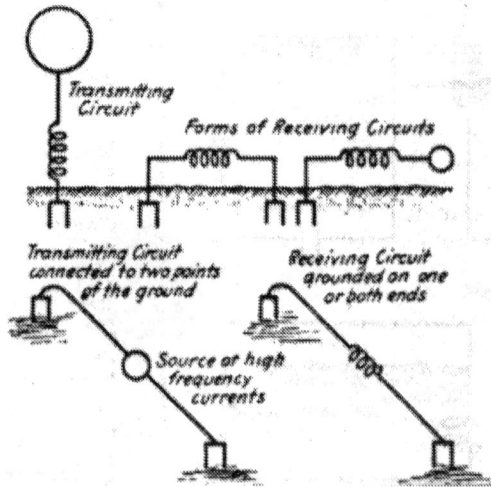

Fig. 12. Arrangements of Directive Circuits Described In Tesla's U. S. Patent No. 613,809 of November 8. 1898, on "Method of and Apparatus for Controlling Mechanism of Moving Vessels or Vehicles."

For the same reason the statics are still the bane of the wireless. There is about as much virtue in the remedies recently proposed as in hair restorers. A small and compact apparatus has been produced which does away entirely with this trouble, at least in plants suitably remodeled.

Nothing is more important in the present phase of development of the wireless art than to dispose of the dominating erroneous ideas. With this object I shall advance a few arguments based on my own observations which prove that Hertz waves have little to do with the results obtained even at small distances.

In Fig. 13 a transmitter is shown radiating space waves of considerable frequency. It is generally believed that these waves pass along the earth's surface and thus affect the receivers. I can hardly think of anything more improbable than this "gliding wave" theory and the conception of the "guided wireless" which are contrary to all laws of action and reaction. Why should these disturbances cling to a conductor where they are counteracted by induced currents, when they can propagate in all other directions unimpeded? The fact is that the radiations of the transmitter passing along the earth's surface are soon extinguished, the height of, the inactive zone indicated in the diagram, being some function of the wave length, the bulk of the waves traversing freely the atmosphere. Terrestrial phenomena which I have noted conclusively show that there is no Heaviside layer, or if it exists, it is of no effect. It certainly would be unfortunate if the human race were thus imprisoned and forever without power to reach out into the depths of space.

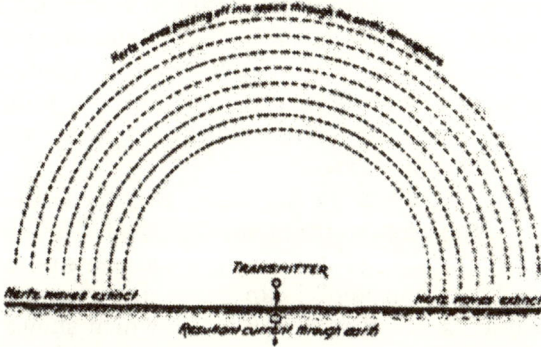

Fig. 13. Diagram Exposing the Fallacy of the Gilding Wave Theory as Propounded In Wireless Text Books.

The actions at a distance cannot be proportionate to the height of the antenna and the current in the same. I shall endeavor to make this clear by reference to diagram in Fig. 14. The elevated terminal charged to a high potential induces an equal and opposite charge in the earth and there are thus Q lines giving an average current $I=4Qn$ which circulates locally and is useless except that it adds to the momentum. A relatively small number of lines q however, go off to great distance and to these corresponds a mean current of $ie = 4qn$ to which is due the action at a distance. The total average current in the antenna is thus $Im = 4Qn + 4qn$ and its intensity is no criterion for the performance. The electric efficiency of the antenna is $q/Q+q$ and this is often a very small fraction.

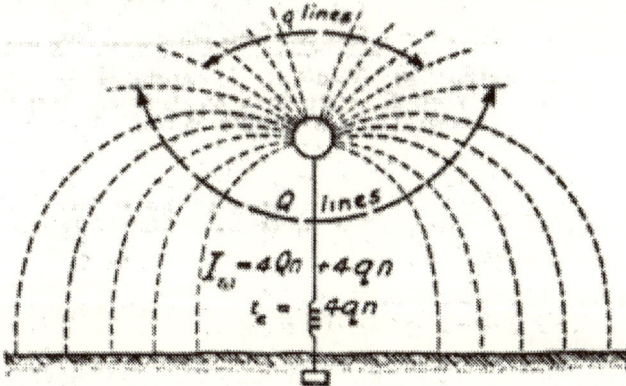

Fig. 14. Diagram Explaining the Relation Between the Effective and the Measured Current In the Antenna.

Dr. L. W. Austin and Mr. J. L. Hogan have made quantitative measurements which are valuable, but far from supporting the Hertz wave theory they are evidences in disproval of the same, as will be easily perceived by taking the above facts into consideration. Dr. Austin's researches are especially useful and instructive and I regret that I cannot agree with him on this subject. I do not think that if his receiver was affected by Hertz waves he could ever establish such relations as he has found, but he would be likely to reach these results if the Hertz waves were in a large part eliminated. At great distance the space waves and the current waves are of equal energy, the former being merely an accompanying manifestation of the latter in accordance with the fundamental teachings of Maxwell.

It occurs to me here to ask the question—why have the Hertz waves, been reduced from the original frequencies to those I have advocated for my system, when in so doing the activity of the transmitting apparatus has been reduced a billion fold? I can invite any expert to perform an experiment such as is illustrated in Fig. 15, which shows the classical Hertz oscillator altho we may have in the Hertz oscillator an activity thousands of times greater, the effect on the receiver is not to be compared to that of the grounded circuit. This shows that in the transmission from an airplane we are merely working thru a condenser, the capacity of which is a function of a logarithmic ratio between the length of the conductor and the distance from the ground. The receiver is affected in exactly the same manner as from an ordinary transmitter, the only difference being that there is a certain modification of the action which can be predetermined from the electrical constants. It is not at all difficult to maintain communication between an airplane and a station on the ground, on the contrary, the feat is very easy.

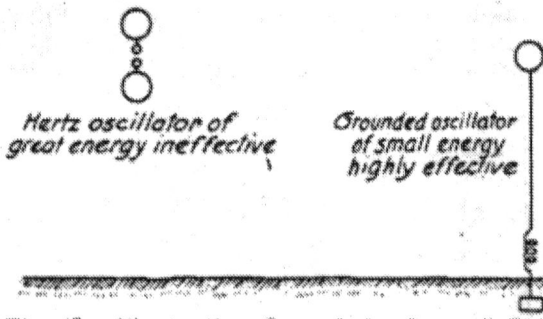

Fig. 15. Illustrating One of the General Evidences Against the Space Wave Transmission.

To mention another experiment in support of my view, I may refer to Fig. 16 in which two grounded circuits are shown excited by oscillations of the Hertzian order. It will be found that the antennas can be put out of parallelism without noticeable change in the action

on the receiver, this proving that it is due to currents propagated thru
the ground and not to space waves.

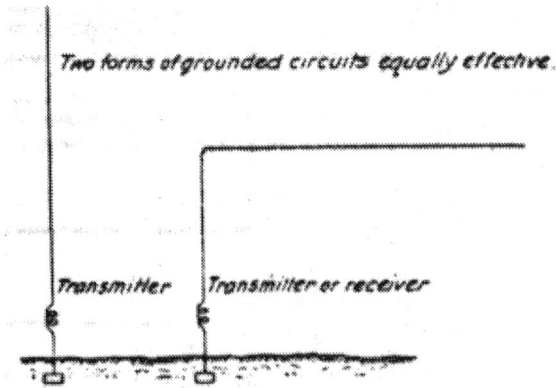

Fig. 16. Showing Unimportance of Relative Position of Transmitting
and Receiving Antennae In Disproval of the Hertz-wave Theory.

Particularly significant are the results obtained in cases illustrated
in Figures 17 and 18. In the former an obstacle is shown in the path of
the waves but unless the receiver is within the effective electrostatic
influence of the mountain range, the signals are not appreciably
weakened by the presence of the latter, because the currents pass
under it and excite the circuit in the same way as if it were attached
to an energized wire. If, as in Fig. 18, a second range happens to be
beyond the receiver, it could only strengthen the Hertz wave effect by
reflection, but as a matter of fact it detracts greatly from the intensity
of the received impulses because the electric niveau between the
mountains is raised, as I have explained with my lightning protector
in the *Experimenter* of February.

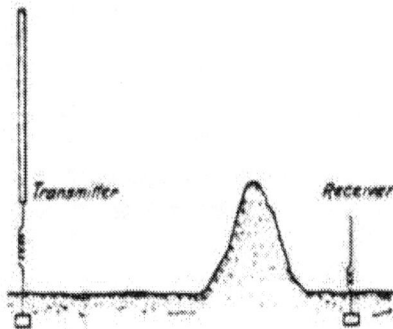

Fig. 17. Illustrating Influence of Obstacle In the Path of Transmission as Evidence Against the Hertz-wave Theory.

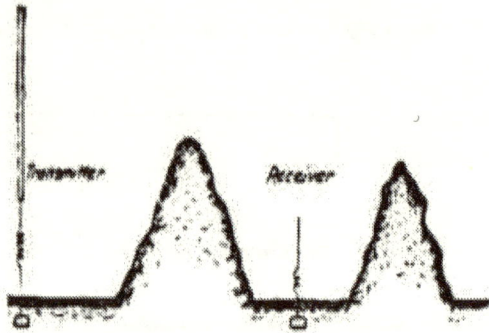

Fig. 18. Showing Effect of Two Hills as Further Proof Against the Hertz-wave Theory.

Again in Fig. 19 two transmitting circuits, one grounded directly and the other thru an air gap are shown. It is a common observation that the former is far more effective, which could not be the case with Hertz radiations. In a like manner if two grounded circuits are observed from day to day the effect is found to increase greatly with the dampness of the ground, and for the same reason also the transmission thru sea-water is more efficient.

Fig. 19. Comparing the Actions of Two Forms of Transmitter as Bearing Out the Fallacy of the Hertz-wave Theory.

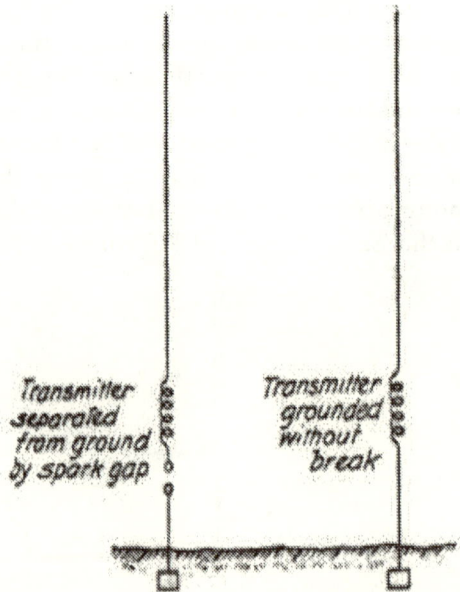

Transmitter separated from ground by spark gap

Transmitter grounded without break

An illuminating experiment is indicated in Fig. 20 in which two grounded transmitters are shown, one with a large and the other with a small terminal capacity. Suppose that the latter be 1/10 of the former but that it is charged to 10 times the potential and let the frequency of the two circuits and therefore the currents in both antennas be exactly the same. The circuit with the smaller capacity will then have 10 times the energy of the other but the effects on the receiver will be in no wise proportionate.

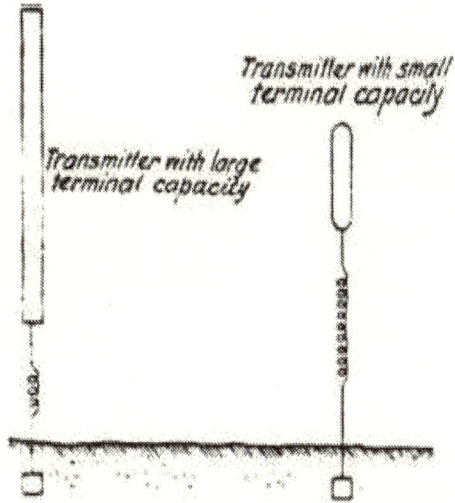

Fig. 20. Disproving the Hertz-wave Theory by Two Transmitters, One of Great and the Other of Small Energy.

The same conclusions will be reached by transmitting and receiving circuits with wires buried underground. In each case the actions carefully investigated will be found to be due to earth currents. Numerous other proofs might be cited which can be easily verified. So for example oscillations of low frequency are ever so much more effective in the transmission which is inconsistent with the prevailing idea. My observations in 1900 and the recent transmissions of signals to very great distances are another emphatic disproval.

The Hertz wave theory of wireless transmission may be kept up for a while, but I do not hesitate to say that in a short time it will be recognized as one of the most remarkable and inexplicable aberrations of the scientific mind which has ever been recorded in history.

www.ingramcontent.com/pod-product-compliance
Lightning Source LLC
Chambersburg PA
CBHW021913040426
42447CB00007B/834